Cyber Foraging

Bridging Mobile and Cloud Computing

Synthesis Lectures on Mobile and Pervasive Computing

Editor
Mahadev Satyanarayanan, *Carnegie Mellon University*

Synthesis Lectures on Mobile and Pervasive Computing is edited by Mahadev Satyanarayanan of Carnegie Mellon University. Mobile computing and pervasive computing represent major evolutionary steps in distributed systems, a line of research and development that dates back to the mid-1970s. Although many basic principles of distributed system design continue to apply, four key constraints of mobility have forced the development of specialized techniques. These include: unpredictable variation in network quality, lowered trust and robustness of mobile elements, limitations on local resources imposed by weight and size constraints, and concern for battery power consumption. Beyond mobile computing lies pervasive (or ubiquitous) computing, whose essence is the creation of environments saturated with computing and communication, yet gracefully integrated with human users. A rich collection of topics lies at the intersections of mobile and pervasive computing with many other areas of computer science.

Cyber Foraging: Bridging Mobile and Cloud Computing
Jason Flinn

ISBN: 978-3-031-01353-9   paperback
ISBN: 978-3-031-02481-8   ebook

DOI 10.1007/978-3-031-02481-8

A Publication in the Springer series
SYNTHESIS LECTURES ON MOBILE AND PERVASIVE COMPUTING

Lecture #10
Series Editor: Mahadev Satyanarayanan, *Carnegie Mellon University*
Series ISSN
Synthesis Lectures on Mobile and Pervasive Computing
Print 1933-9011   Electronic 1933-902X

Cyber Foraging

Bridging Mobile and Cloud Computing

Jason Flinn
University of Michigan

SYNTHESIS LECTURES ON MOBILE AND PERVASIVE COMPUTING #10

ABSTRACT

This lecture provides an introduction to cyber foraging, a topic that lies at the intersection of mobile and cloud computing. Cyber foraging dynamically augments the computing resources of mobile computers by opportunistically exploiting fixed computing infrastructure in the surrounding environment. In a cyber foraging system, applications functionality is dynamically partitioned between the mobile computer and infrastructure servers that store data and execute computation on behalf of mobile users. The location of application functionality changes in response to user mobility, platform characteristics, and variation in resources such as network bandwidth and CPU load. Cyber foraging also introduces a new, *surrogate* computing tier that lies between mobile users and cloud data centers. Surrogates are wired, infrastructure servers that offer much greater computing resources than those offered by small, battery-powered mobile devices. Surrogates are geographically distributed to be as close as possible to mobile computers so that they can provide substantially better response time to network requests than that provided by servers in cloud data centers. For instance, surrogates may be co-located with wireless hotspots in coffee shops, airport lounges, and other public locations.

This lecture first describes how cyber foraging systems dynamically partition data and computation. It shows how dynamic partitioning can often yield better performance, energy efficiency, and application quality than static thin-client or thick-client approaches for dividing functionality between cloud and mobile computers.

The lecture then describes the design of the surrogate computing tier. It shows how strong isolation can enable third-party computers to host computation and store data on behalf of nearby mobile devices. It then describes how surrogates can provide reasonable security and privacy guarantees to the mobile computers that use them. The lecture concludes with a discussion of data staging, in which surrogates temporarily store data in transit between cloud servers and mobile computers in order to improve transfer bandwidth and energy efficiency.

KEYWORDS

cyber foraging, mobile code offload, dynamic partitioning of computation, data staging

Contents

CHAPTER 1

Introduction

1.1 THE MOTIVATION FOR CYBER FORAGING

Mobile applications increasingly pervade our daily lives. Innovation in computing hardware has continually provided mobile users with more computational resources in lighter and smaller form factors, as laptops have replaced personal computers, and are now in turn being replaced by tablets and smart phones. Hardware advances have in turn triggered a deluge of new, mobile software applications; the mobile application market is expected to reach nearly $30 billion by 2013 [Gartner, 2010]. The success of this market is driven by user demand for pervasive access to data and computation. The portability of current mobile devices satisfies this demand by allowing their users to run applications anywhere at any time.

Yet, most mobile applications do not run in isolation on the mobile device. Instead, they access data and computational resources in the cloud via wireless networking infrastructure such as cellular base stations and WiFi access points. Wired infrastructure components such as compute servers and data stores provide more computational resources than a mobile device because they are not limited by stringent size, weight, and battery energy constraints. This combination of mobile devices and cloud infrastructure is so compelling that technologists have long posited the "post-PC era" in which mobile devices replace desktops and workstations as the primary way in which users interact with their applications and data.

The main question examined by this lecture is: *How should one best partition applications across mobile platforms and fixed infrastructure*? Current applications typically address this question in an ad-hoc fashion, either by operating in a *thin-client* mode in which all computation and data access is performed by the fixed infrastructure and the mobile device operates only as a terminal or by operating in a *thick-client* mode in which the computation and data are hosted on the mobile device and wireless networks are used only to fetch inputs not available locally. A handful of applications adopt a more balanced, but still static, partitioning of functionality.

Static partitioning, however, fails to account for the variability inherent in mobile and pervasive computing environments. Wireless network quality can change by one or more orders of magnitude as users move. Mobile devices may become completely disconnected from the fixed infrastructure for periods of time. Shared computational resources in the cloud may become overtaxed. Nearby computational resources may become available for opportunistic usage. The resources available to different mobile devices can vary substantially from platform to platform. Battery energy and cellular bandwidth are limited resources and may need to be consumed within a budget. Application demand and user workloads may change over time. The cumulative effect of this variability is that it is infeasible to select an "always-best" static partitioning of application functionality. Since there is no single scenario in which a mobile application is employed, any static partitioning will inevitably lead to a poor user experience when the actual usage environment varies from that envisioned by the developer.

Cyber foraging represents a different approach to building mobile applications, in which the partitioning of application functionality between mobile and infrastructure platforms is fluid. As originally envisioned, cyber foraging "dynamically augments the computing resources of a wireless mobile computer by exploiting wired hardware infrastructure." [Satyanarayanan, 2001]. Cyber foraging enables the seamless mobility of data and computation. Users are given the illusion that their applications run entirely locally on their mobile devices; this illusion preserves the pervasive, always-available access to applications and data that has driven the growth of mobile computing. Behind this illusion, however, computation and data may be replicated and migrated between the mobile computer and fixed infrastructure. This infrastructure may either be designated resources in the cloud or nearby compute and storage resources that are dynamically identified and opportunistically exploited. As the conditions of the mobile environment changes, cyber foraging systems dynamically adjust the partitioning of computation and data to provide the best user experience. Thus, as long as sufficient infrastructure resources can be harnessed from the environment, cyber foraging can provide mobile users with the performance they have come to expect from better-equipped computing platforms such as desktop computers.

We next provide a more detailed overview of the potential benefits and costs of exploiting remote infrastructure. During this overview, we note that many costs and benefits are dependent on the environment in which the mobile application is executed; this situational-dependence is the key element that drives the need for dynamic partitioning. We conclude the introduction with a road map for the remainder of the lecture.

1.2   POTENTIAL BENEFITS FROM USING REMOTE INFRASTRUCTURE

There are many reasons why an application may benefit from executing a computation or storing data on remote infrastructure instead of or in addition to the mobile computer on which it is currently executing. The most obvious potential benefit is performance. Since mobile users value portability, mobile computers such as smart phones and tablets must be much smaller and lighter than server and desktop computers. These constraints create a fundamental performance gap between mobile and fixed computers — even as computing power has increased rapidly over the past few decades, the computational resources available to mobile devices have substantially lagged behind their fixed counterparts.

Figure 1.1 illustrates this trend by showing the processing power of representative server and handheld computers over the past 15 years. For each five-year increment, the second and third columns show the processor speed of a typical server using an Intel processor sold during that year. The final two columns show a mobile, handheld computer released during that year and the speed

Year	Representative server		Representative handheld	
	Processor	Speed	Device	Speed
1997	Pentium II	266 MHz	PalmPilot	16 MHz
2002	Itanium	1 GHz	Blackberry 5810	133 MHz
2007	Core 2	9.6 GHz (4 cores)	Apple iPhone	412 MHz
2012	Xeon E3	14 GHz (2x4 cores)	Samsung Galaxy 3S	3.2 GHz (2 cores)

This figure compares the processing power of representative computer systems in five-year increments from 1997–2012. As a rough estimate, processing power is given by the product of core count and clock speed. Although both server and mobile computers both show rapid growth in processing power, the performance gap between the two remains substantial for every time period examined.

Figure 1.1: Comparison of mobile and infrastructure processing power.

of its processor. For computers with multiple cores, the table calculates speed as the number of total cores multiplied by the individual processor speed (this is only intended to be a rough estimate of processing power). Both server and handheld computers have seen remarkable growth in processing power, first through increasing processor clock speed and later due to increased parallelism realized via multicore chips. However, in each year, there is a substantial performance gap between the capabilities of server and handheld computers. The gap is lowest for modern computers (2012), but these results due to not account for distributed systems in cloud data centers that can leverage many servers to perform compute intensive tasks.

Due to the gap between mobile and infrastructure processing power, compute-intensive applications can execute much faster on remote infrastructure than on mobile devices. On the other hand, interactive activities that demand few computational resources may execute almost as fast on a mobile computer as they do on a server. Further, performance is not impacted solely by processor speed; remote infrastructure may offer more memory and storage, the ability to parallelize computation across multiple cores and servers, or better network connectivity.

The second potential benefit is reducing energy usage. A mobile computer system operates on battery power. It must budget this finite source of energy wisely so that it can perform all the activities demanded by its users without exhausting the supply before an opportunity to recharge the battery arises. Designers therefore strive to make mobile computing systems as energy-efficient as possible, for example, by employing hardware power-saving modes or by reducing the scope or quality of activities performed by mobile applications. While these measures are essential to extending battery lifetime, they also noticeably degrade the mobile user experience; applications take longer to perform interactive activities and produce lower-quality results. Use of fixed infrastructure is an attractive alternative for designers. By offloading computation or data storage from a battery-powered computer to a remote computer with wall power, the operational lifetime of the battery-powered, mobile computer can sometimes be extended without degrading the user experience.

A third potential benefit is increasing application *fidelity*, which we define as the degree to which the results produced by an application match the highest quality results that would be produced given ample computational resources. We have already discussed two reasons why a mobile computer operating alone may choose to reduce application fidelity. First, reducing fidelity may be required to achieve acceptable response times for interactive applications. Second, reducing fidelity can extend the mobile computer's battery lifetime. While these reasons represent explicit design choices, for demanding applications, fidelity reduction may sometimes be absolutely necessary in order to get the application to run on the mobile device at all. For instance, a fixed computer may be able to execute a intense computation that a mobile computer is simply incapable of doing locally due to insufficient memory or storage. Offloading a computation or data query to a cluster in the cloud provides access to orders of magnitude more resource than a mobile computer has available locally. In such a situation, the mobile system designer has only two choices: execute the application locally with reduced fidelity, or leverage fixed infrastructure to provide full fidelity.

1.3   POTENTIAL COSTS OF USING REMOTE INFRASTRUCTURE

Unfortunately, the use of fixed infrastructure does not always come without cost. In many circumstances, offloading application activity from the mobile computer may actually degrade performance. For instance, in order to execute a computation remotely, the mobile computer must ship the inputs to that computation to the remote infrastructure over a wireless network. After the computation completes, the mobile computer must retrieve the results of the computation. Unless the computation is asynchronous and not on the critical path of the application, performance is improved only if the time saved by performing the computation remotely is greater than the time spent communicating inputs and outputs at the start and end of the computation. When network latency between the mobile and remote computers is high, when bandwidth is low, or when the amount of data shipped per unit of computation is large, it may be quicker to perform the computation on the mobile computer than it would be to perform the computation remotely.

The equation for energy savings is similar. While the mobile computer may be able to use aggressive hardware power-saving modes while application activity is performed remotely (saving energy), it must use a power-hungry wireless network to initiate the remote activity and receive the results (consuming energy). Under the same circumstances outlined for performance, it may require more energy to offload activity to a remote computer than it would require to do the same activity locally.

Offloading application activity to remote infrastructure introduces security and privacy concerns. For instance, an attacker with physical access to the infrastructure may cause an application to return incorrect results, or an eavesdropper may monitor the remote computation to discover private data. Of course, in an era in which mobile users routinely download and run third-party applications, hosting computation solely on a mobile device is no guarantee of either privacy or security (in fact, a recent study found that many popular Android applications leak sensitive information

Factor	Favorable for remote execution	Favorable for local execution
Network bandwidth	High	Low
Network latency	Low	High
Disparity between compute resources	High	Low
Granularity of computation	Large	Small
Parallelism in computation	High	Low
Memory/storage requirements	High	Low
Size of inputs/outputs	Small	Large
Load on remote computers	Low	High
Wireless network power	Low	High
Sensitivity of activity	N/A	Sensitive

Figure 1.2: Some factors that impact the partitioning decision.

to remote sources [Enck et al., 2010]). Yet, physical access to remote hardware opens an additional attack vector not present when applications are confined solely to a personal device.

1.4 DISCUSSION

Figure 1.2 extracts from the previous discussion a list of some of the factors that potentially impact the partitioning decision, i.e., the decision about whether to execute application activity locally on the mobile computer or remotely on fixed infrastructure. While this list is incomplete (the remainder of the lecture will cover additional decision criteria for the partitioning decision), even a cursory consideration of the list yields several important conclusions. First, the partitioning decision is complex: a cyber foraging system must not only balance numerous goals (e.g., performance, energy-efficiency, fidelity, privacy, and security), but it must also evaluate a large number of factors including those in Figure 1.2 that impact those goals. Second, all of the factors listed may need to be evaluated dynamically. Network conditions and load may change rapidly in mobile environments. Factors such as the granularity of computation and the sensitivity of data may depend upon specific inputs chosen by the user at runtime. There is simply no "one-size-fits-all" choice for all of these factors. Hence, it is infeasible to expect that a static partitioning decision can yield optimal or even good results in many environments in which a mobile computer may operate. Achieving good results requires the dynamic decision making of a cyber foraging system.

1.5 LECTURE OVERVIEW

In this Introduction we discussed the advantages and disadvantages of offloading computation and data storage from a mobile computer to fixed infrastructure. We have shown that the criteria for this decision vary substantially in mobile environments, creating the need for cyber foraging systems that make partitioning decisions dynamically rather than statically. The following chapters examine the design of cyber foraging systems in greater depth.

Chapter 2 examines the algorithms used by current cyber foraging systems to make decisions that dynamically partition applications. It concludes by detailing some of the open research questions in this area.

Chapter 3 examines issues with managing remote infrastructure, including discovery, isolation, ease-of-use, and load management.

Chapter 4 examines security and privacy issues introduced by cyber foraging, detailing new threat models and attacks, as well as current mechanisms used to mitigate those attacks.

Chapter 5 examines storage offloading, including mechanisms for populating remote stores and algorithms for deciding how to partition data between the mobile and remote systems.

Chapter 6 looks forward by examining current challenges and opportunities for cyber foraging systems, including the business case for deploying remote cyber foraging infrastructure, technological barriers, and opportunities for future research.

CHAPTER 2

Partitioning

This chapter examines the partitioning decision that lies at the heart of cyber foraging. In essence, the partitioning decision answers the following question: *Given a specific application state and a specific computational environment, which portions of the application should run on the mobile computer and which should run on remote infrastructure?* The dynamic nature of this question arises from two sources. First, the application state may depend on specific inputs and the current location of data used in the computation. Second, the mobile environment (server load, network bandwidth, etc.) is constantly changing.

Current cyber foraging systems use a range of approaches to place application components. To better discuss the similarities and differences among the diverse approaches, we decompose the partitioning decision down into smaller sub-problems. First, in Section 2.1, we examine the different objectives that cyber foraging systems attempt to achieve through application partitioning and the metrics used to quantify those objectives. In Section 2.2, we describe different approaches to enumerating the set of candidate partitions. Section 2.3 discusses strategies used to select one of those candidates. Most selection strategies require estimates of resource supply and demand, so Section 2.4 explains how current cyber foraging systems derive such estimates. Section 2.5 describes how cyber foraging recovers from failures that occur during partitioned execution. Section 2.6 lists the applications that have been partitioned by cyber foraging systems and reflects on what characteristics make applications particularly well suited or poorly suited for cyber foraging. Section 2.7 concludes with a summary of the chapter.

2.1 PARTITIONING GOALS

All systems that partition application components among mobile computers and fixed infrastructure attempt to place application components such that some metric is maximized. Sections 1.2 and 1.3 outlined the metrics that are commonly used to guide the partitioning decision. However, there is certainly no consensus among cyber foraging systems about which of these metrics to employ; most systems only consider a subset of them when deciding where to place application components.

The Remote Processing Framework [Rudenko et al., 1998], one of the earliest systems to support dynamic partitioning of functionality among mobile computers and fixed infrastructure, attempts only to minimize the energy consumed by the mobile computer (so as to extend its battery lifetime). CloneCloud [Chun et al., 2011], a very recent cyber foraging system, may minimize either mobile computer energy usage or application execution time.

Cyber Foraging System	Goals	Method of Relating Goals
RPF [Rudenko et al., 1998]	Energy	N/A
Spectra [Flinn et al., 2002]	Execution time, energy, and fidelity	Utility function and goal-directed adaptation
Chroma [Balan et al., 2003]	Execution time and fidelity	Utility function
MAUI [Cuervo et al., 2010]	Execution time and energy	Optimize energy subject to execution time constraint
CloneCloud [Chun et al., 2011]	Execution time or energy	N/A
Odessa [Ra et al., 2011]	Makespan and throughput	Only execute remotely if both goals improve

This table summarizes the partitioning goals of the cyber foraging systems discussed in this chapter. The second column shows the goals employed by each system and the third column show the method used to relate goals expressed in different metrics.

Figure 2.1: Partitioning goals for different cyber foraging systems.

Other systems consider multiple goals simultaneously. Odessa [Ra et al., 2011] considers two different performance goals: throughput, which is the rate at which operations can be performed, and makespan, which is the difference between start and end times for a series of operations. MAUI [Cuervo et al., 2010] considers both energy and execution time. Chroma [Balan et al., 2003] weighs both execution time and fidelity (application quality) when making decisions. Spectra [Flinn et al., 2002] balances execution time, energy, and fidelity concerns. The second column of Figure 2.1 provides a summary the goals chosen by the designers of these cyber foraging systems.

Interestingly, no system currently considers security or privacy explicitly when making dynamic partitioning decisions. One potential reason for this choice may be that is hard to determine which data is sensitive and which is not without specific user annotations [Yumerefendi et al., 2007]. Further, the negative consequences of making incorrect security and privacy decisions can be large. Unfortunately, this means that all security concerns must currently be addressed statically, perhaps limiting the set of applications that can benefit from cyber foraging.

Systems that attempt to optimize more than one metric face an additional challenge: there is no common currency for these metrics. For instance, energy expenditure is expressed in Joules, performance in seconds, and fidelity is an abstract notion of application quality. When a particular partitioning optimizes all metrics, that partitioning is the obvious choice. When no such partitioning exists, the cyber foraging system must sacrifice one metric to improve another.

One method for optimizing multiple metrics is to treat all metrics save one as constraints and optimize the remaining metric subject to those constraints. MAUI uses this method to make its partitioning decisions. Its solver minimizes energy usage on the mobile computer subject to the constraint that performance is no worse than 105% of the performance that would be achieved by executing the activity entirely on the mobile device.

An alternative method is to execute functionality only if all performance metrics indicate that remote execution is favorable. Odessa uses this approach: it only executes application components on remote infrastructure if it expects to improve both throughput and makespan as a result.

A third possible method defines conversion factors between metrics. For instance, Chroma's default behavior assigns equal importance to fidelity and execution time [Balan et al., 2003]. However, Chroma also provides an interface to specify a *utility function* that expresses arbitrary relationships or constraints among metrics. Spectra uses a similar approach to balance energy, execution time, and fidelity.

Each of the above methods represents a type of policy for relating metrics expressed in different currencies. The third column of Figure 2.1 summarizes the choices made by each of the cyber foraging systems discussed in this chapter.

Unfortunately, the selection of the actual policy used is often arbitrary: it may or may not match the application user's preferences. If a system developer chooses a poor policy, then the cyber foraging system may degrade the user's experience.

A few methods attempt to solicit user input to help set the policy dynamically. One approach, used by Spectra, is based on the observation that the relative importance of metrics may change over time. Spectra asks its user to specify a target battery lifetime for the mobile device at runtime. It uses a feedback process termed *goal-directed adaptation* [Flinn and Satyanarayanan, 2004] to dynamically adjust the relative importance of energy in relation to fidelity and execution time based on how well the rate at which the mobile computer is consuming power energy matches the target battery lifetime. When the mobile computer is projected to run out of battery energy too soon, Spectra increases the relative importance of energy-efficiency. If the mobile computer is projected to have substantial energy left over, Spectra decreases the relative importance of energy efficiency. A similar approach could potentially be used to address other *budgeted resources* for which a fixed maximum consumption is allowed over a time period. For instance, goal-directed adaptation has also been used to ensure that a mobile phone does not exceed a monthly cap on a cellular data plan [Higgins et al., 2012].

The Aura project [Sousa et al., 2008] uses off-line profiling to construct utility functions. This approach allows users to express thresholds for satiation and starvation for multiple metrics. It then interpolates among those specifications using sigmoid functions.

In general, these systems illustrate that there is an inherent tradeoff in soliciting user input. More user participation can improve the quality of the partitioning, but the participation may distract the user from the task at hand and prove too burdensome.

2.2   CHOOSING CANDIDATE PARTITIONS

After choosing metrics and constraints that express the goals for partitioning, the designer of a cyber foraging system must next decide on a method for determining *candidate partitions* for applications. Theoretically, the number of possible application partitions is very large since partitioning could employ granularities as fine as single instructions. From a practical standpoint, however, very fine-grained partitions are ineffective because substantial computation is needed to offset the performance and energy costs of using the network. Further, the performance overhead of the partitioning decision must be small since it is executed dynamically; this limits the number of possible partitions that can be considered. For these reasons, cyber foraging systems typically first enumerate a small number of possible partitions, which we call the candidate partitions, and then dynamically select the candidate partition that best achieves the system goal.

There is considerable difference of opinion about how best to enumerate candidate partitions. One approach, as exemplified by Chroma, is to ask a programmer to specify the possible ways of partitioning an application. This approach is based on the belief that "for every application, the number of useful ways of splitting the application for remote execution is small." [Balan et al., 2003] Because the number of candidate partitions is small, the granularity of partitioning is typically large application components. For instance, when adapting the Pangloss-Lite natural language translator [Frederking and Brown, 1996] to their system, the Chroma developers identified four components for partitioning (three translation engines and a language modeler) and enumerated seven candidate partitions among those four components.

The developers of Chroma subsequently created a little language called Vivendi Balan et al. [2007] that allows developers to specify a compact static partition of all meaningful partitions of an applications. Using Vivendi, an application developer specifies code components called *remoteops* that may benefit from remote execution given the right circumstances. The developer also specifies *tactics*, each of which is a different way of combining remote procedure calls to produce a remoteop result. At runtime, the Chroma system selects one of the specified tactics.

Many other systems follow this approach. Like Chroma, Spectra asks the developer to specify candidate partitions explicitly. RPF uses a more extreme version of this approach, in which it considers only two candidate partitions: one in which the core computation is performed on the mobile computer and one in which the computation is performed remotely.

The other extreme is to choose candidate partitions at method granularity without requiring any programmer assistance. This approach takes advantage of modern language runtimes to identify application components that can be remotely executed and the data on which those components operate.

CloneCloud [Chun et al., 2011] offers the best example of this approach. It identifies candidate partitions through static analysis of code compiled to run in Android's Dalvik virtual machine. CloneCloud prunes the set of candidate partitions by considering only those partitions that start and end at VM-layer method boundaries. It imposes three additional restrictions. First, methods that access platform-specific features (such as the GPS device on a mobile phone) must execute on the platform that has those features. Second, methods that access the same native state (e.g., a temporary file) must execute on the same computer. Finally, nested migration is disallowed — if a method is migrated from the mobile computer to remote infrastructure, the methods it invokes must also execute remotely. While one can imagine removing each of these restrictions (e.g., by supporting

Cyber Foraging System	Partition Granularity	Programmer Effort
RPF	Whole application	None
Spectra	Programmer-specified	Identify components
Chroma	Programmer-specified	Identify components
MAUI	Method	Annotate remotable methods
CloneCloud	Method	None
Odessa	Sprout component	None, but must use Sprout

This table summarizes the strategies employed by different cyber foraging systems for identifying candidate partitions. The second column shows the minimum granularity of a component that can be executed remotely. The third column shows the programmer effort required to identify candidate partitions.

Figure 2.2: Methods for identifying candidate partitions.

remote service invocation or migration of native state), the restrictions support the common case while substantially reducing the set of candidate partitions that CloneCloud needs to consider.

Other systems use a hybrid of these two approaches. For example, MAUI also considers candidate partitions specified at method granularity. However, MAUI asks developers to specify which methods it should consider executing remotely by annotating those methods with the custom attribute feature of the Microsoft .NET Common Language Runtime. While this requires that the developer perform some of the checks that are done automatically by CloneCloud (for example, verifying that the method does not access functionality specific to the mobile platform), it also allows the developer to pass along domain-specific knowledge (for instance, that it seldom is a good idea to execute user interface methods remotely).

Language runtimes represent only one way for a cyber foraging system to automatically extract application components. If an application has already been divided into components to use distributed systems middleware, then the cyber foraging system can leverage the same component structure to identify candidate partitions. For example, Odessa targets applications that have already been modified to use the Sprout [Pillai et al., 2009] distributed stream processing system. Sprout models applications as a data flow graph in which the vertices are computational components called stages; Odessa dynamically decides where stages should be placed and when to employ parallel computation. Although not targeted specifically at mobile computing, Coign [Hunt and Scott, 1999] uses a similar strategy: by targeting applications built using Microsoft's Common Object Model (COM), Coign can learn of partition boundaries and distribute an application without any knowledge of the application source code.

Figure 2.2 summarizes the different approaches employed by cyber foraging systems to identify candidate partitions. Despite the diversity of approaches, experimental results from these different systems exhibit substantial similarity. We can extract several useful conclusions from these similar results.

First, Chroma and similar systems hypothesize that there are a small number of useful ways to partition an application. Results from CloneCloud and MAUI validate this hypothesis. Even though these systems consider a large number of candidate partitions, only a few candidates are actually employed in practice. For all applications considered, CloneCloud either offloads the core computation to remote infrastructure or executes the application entirely locally. MAUI uses two partitions in practice for most applications, and it uses three for a video game.

Why is the number of useful partitions small? One reason may be that it only makes sense to offload application components that contain substantial computation, and there are only one or two such components in typical applications. This hypothesis indicates that if network throughput and latency were to dramatically improve, the number of components that could be offloaded profitably might increase.

Even though the number of useful partitions is small for most applications, it might be the case that automated methods such as those employed by CloneCloud, Coign, and MAUI are better able to identify those partitions than human programmers. Balan et al. [Balan et al., 2007] report that the applications they modified for Chroma were already structured to make decomposition easy, but not all applications are well structured.

Developer effort may be a more important consideration. Systems such as CloneCloud and Coign can automatically identify candidate partitions without any need to modify the application. Potentially, these systems can partition any application written for their target runtimes, which dramatically lowers the barrier to employing cyber foraging in practice. The flip side of this is a sacrifice of generality: applications not written to use the target runtime cannot be handled. Of course, on platforms such as Android phones for which the majority of applications use the target runtime, this may not be a substantial issue.

One final drawback of method granularity is the computational cost of choosing from among a much larger number of candidate partitions. Systems such as Chroma exhaustively iterate through candidates and choose the one that maximizes the chosen partitioning metric; this process is quick since the set of candidates is small. With a much larger set of candidates, MAUI employs a linear program solver to choose the best candidate — this takes 18–46 ms, depending on application complexity. Since executing the solver at every potential offload point would be prohibitively expensive, MAUI runs the solver periodically in the background and caches the results. CloneCloud also uses an optimization solver. To remove the solver from the critical path, CloneCloud precomputes the optimal partitions for different combinations of network, CPU, and energy conditions. It then chooses an appropriate partition from among the precomputed ones based on observed conditions when the application runs.

2.3   SELECTING A PARTITION

Given a set of candidate partitions, a cyber foraging system selects the one that it believes will best achieve its goals. As discussed in Section 2.1, this should be the partition that maximizes the system's chosen metric or utility function while satisfying any constraints. Unfortunately, this choice is necessarily imperfect because making it correctly requires the cyber foraging system to predict the future. An accurate prediction requires the system to estimate not only the conditions of the mobile computing environment (for example, network bandwidth, latency, and server load) but also application behavior (for instance, the amount of computation that will be performed by a component for a specific set of inputs). Typically, the cyber foraging system assumes that the past will predict the future. It observes application behavior and/or the resources available to the mobile computer and bases future predictions on those observations.

There is a tradeoff between simplicity and accuracy in estimating the benefit of each partition. RPF is an example of a system that opts for simplicity. RPF uses an estimation method that we will term *direct observation*. RPF's chosen metric is mobile computer energy usage, and it considers only two partitions. Initially, RPF requires the application to alternately execute each partition. Every time the application executes, RPF directly measures the drain on the mobile computer's battery through the ACPI interface [Intel et al., 1998]. It uses a digital filter to produce a smoothed estimate of the energy requirements of each partition. After it has gathered sufficient measurement, RPF uses the partition with the lowest estimated energy cost most of the time, while still executing the other partition occasionally to see if its energy costs have changed.

While the direct observation approach is appealing in its simplicity, it does have some drawbacks. First, the decision engine does not scale well with the number of partitions. The decision strategy requires that sufficient measurements be gathered for all partitions before RPF can start to choose the best partition, and it also requires that all partitions be periodically sampled to verify that their costs have not changed.

Another drawback of direct observation is that it only considers the observable effects of executing a particular partition. It does not try to understand why the partition produced the measured result. This can be disadvantageous when the computational and data requirements of an application exhibit substantial variation depending on inputs. In that case, an application might exhibit high energy usage not because it is executed using the wrong partition, but rather due to the specific inputs with which it was executed. If the application is not executed frequently, past measurements might be stale and may not be a good prediction for the cost of executing each partition in the current environment. Finally, each prediction is application-specific: RPF cannot use measurements from one application to benefit another.

These drawbacks have led other cyber foraging systems to adopt more nuanced prediction methodologies that separately predict resource *supply* (the resources available in the mobile computer's computational environment) and resource *demand* (the requirements of each candidate partition). For ease of discussion, we will describe systems that use this *supply and demand* method starting with the simplest approach and following with more complex refinements. We note, however, that this does not reflect the chronological order in which the systems were developed.

MAUI uses a straightforward application of the supply and demand method to estimate the time spent transferring state over the network when a component is executed remotely. The demand in this case is the amount of bytes required to send the inputs for the computation to the remote server plus the amount of bytes required to ship the results of the computation back to the mobile computer. MAUI estimates network demand through direct observation. When any method A calls a method B, it measures the state that would need to be transferred over the network to enable B to run remotely.

MAUI uses measurements of past demand are used to predict future demand. It actively measures network characteristics (bandwidth, latency, and packet loss) to estimate the future supply of the resource (again, the recent past is assumed to be a good predictor of the future). By combining these estimates of network supply and demand, MAUI predicts how much time will be spent transferring state if it chooses to execute a method remotely.

The advantage of the supply-and-demand approach is that it can quickly adapt its predictions in response to changes in the mobile computing environment. For example, when MAUI observes a increase in network bandwidth or a decrease in latency, it can immediately update its predictions for network transfer times for *all* methods. With a supply-and-demand approach, a system like RPF needs to observe the effect of the change in network behavior by executing a partition that includes a remote component. Due to smoothing of observations, it may need multiple observations to change its behavior. Finally, if there are multiple components that can be executed remotely, the system must directly observe the effect of the change in network behavior on each individual component before it can change its partitioning decision for that component.

While the approach used by MAUI can adapt quickly to changes in supply (i.e., network conditions), it is less agile in adapting to changes in demand. Often, the amount of state that must be transferred between the mobile and remote computers is input-dependent. When inputs vary, the size of the state transfer will change in response. However, MAUI must observe the change in state size directly before it adjusts its partitioning decision. This limitation is similar to that observed for RPF, and results from demand being estimated through direct observation.

One of the earliest systems to use the supply and demand method [Narayanan et al., 2000] addressed this limitation with a technique termed *history-based prediction*. This is a empirical methodology that combines profiling and online passive observation to learn an application's behavior and predict its resource consumption. History-based prediction was initially targeted at *multi-fidelity computation*, in which application quality is adjusted to meet resource constraints. However, later work applied this technique to remote execution [Narayanan, 2002], and it has been used in cyber foraging systems such as Spectra and Chroma.

When history-based prediction is used, the application is first sampled to explore its behavior under different inputs when performing computation at different fidelities. Based on the profile data, machine learning algorithms are used to predict how application resource usage varies with fidelity and inputs. These predictions are used when the application runs; however, the model is also updated based on new passive measurements of application resource usage.

The need for a profiling phase is one of the drawbacks of history-based observation. However, it may be a necessary evil that is required to explore the space of possible options. The alternative, used in RPF, is to force the application to explore the space as it is used; this option is potentially distracting to the user when poor choices are being explored.

History-based prediction also requires some hints from the application developer. In particular, the developer must specify the inputs to a computation, the fidelities at which the computation may be performed (if applicable), and the algorithmic complexity of the computation with respect to the usage of a particular resource (e.g., CPU or memory). These hints are provided via an Application Configuration File (ACF). For example, a developer might specify that the running time of a computation varies linearly with one input and quadratically with a second input.

Narayanan chose to use exponentially weighted recursive least-squares estimation (RLS) [Young, 1984] as the machine learning algorithm for history-based prediction. This choice balances two objectives. First, one wants to provide the best fit over the sampled data. Least-squares regression is an attractive methodology for this purpose (as long as all non-linear relationships are specified in the ACF). Second, one also wants to adjust for changes in application behavior over time. RLS achieves both purposes as it is a simple refinement to least-squares regression that gives more weight to recent samples. Essentially, it is a combination of least-squares regression and exponentially weighted moving average predictors.

Putting it all together, the profile data serves as the initial starting point for the RLS estimator. The estimator is updated with more recent estimates collected as the application runs. Narayanan et al. also describe a caching optimization [Narayanan et al., 2000] in which the resource usage of previously seen inputs are remembered and provided by the predictor if a given input is seen more than once.

History-based prediction is thus resilient to changes in application input. Even when resource demand varies considerably from one execution to another, a history-based predictor can accurately learn the demand as a function of those inputs. However, the cost of this accuracy is additional work on the part of the application developer: specification of inputs and algorithmic complexity, as well as profiling application behavior over a representative set of inputs.

There is considerable room for innovation in the space of algorithms that select among candidate partitions. Odessa uses a novel approach in which it first attempts to identify the performance bottleneck in an application using a lightweight application profiler. Its decision engine uses simple predictors based on processor frequencies and recent network measurements to estimate whether offloading a component from the mobile computer to a remote server or increasing the level of parallelism for processing the component would improve performance. This is a greedy, incremental approach, but results from several parallel applications show that the approach works well in practice.

MARS [Cidon et al., 2011] uses a similar greedy, incremental strategy for deciding which components to run remotely. It sorts tasks by the ratio of predicted local execution time to predicted remote execution time. It offloads components with the highest such ratio first, subject to an energy constraint. This strategy does not adjust for inter-component dependencies and thus may be best suited for applications that exhibit high degrees of potential parallelism.

Cyber Foraging System	Cost/Benefit Estimation	Resources Measured
RPF	Direct observation	None
Spectra	History-based prediction	CPU, network, energy, and cache state
Chroma	History-based prediction	CPU, memory, network, energy, and cache state
MAUI	Supply and demand	CPU, network, and energy
CloneCloud	Supply and demand	Network
Odessa	Bottleneck identification	Network and remote parallelism

This table summarizes the strategies employed by different cyber foraging systems to choose among candidate partitions. The second column shows the method used to identify the best candidate and the third column summarizes the resources measured to make that identification.

Figure 2.3: Methods for selecting among candidate partitions.

Figure 2.3 summarizes the different strategies used to identify candidate partitions. Other than direct observation, all of the above methods require that the cyber foraging system estimate the supply and/or demand of one or more resources. However, each resource is a bit unique, and each has its own peculiarities for measurement and estimation. Hence, slightly different techniques are used for each resource. The next section describes these techniques for each of the resources that impact the partitioning decision.

2.4   RESOURCE MEASUREMENT AND ESTIMATION

The resources that cyber foraging systems most often use for supply-and-demand estimation are CPU, network, battery energy, file cache state, and memory. We first explore how each of these resources are measured and estimated, and then we speculate on what other resources may potentially be of interest to cyber foraging systems.

2.4.1   CPU

Most operating system kernels maintain per-application CPU usage statistics. Cyber foraging systems such as Spectra and Chroma measure the demand (usage) of CPU for an application by reading these statistics through interfaces such as Linux's /proc file system. The per-application nature of the statistics enables the cyber foraging system to distinguish the usage of the target application from other concurrent activities on the computer, although some conflating effects (such as lock and cache contention) are very difficult to remove. Similarly, cyber foraging systems often use interfaces such as the /proc file system to measure supply by querying the current CPU load for the computer.

One challenge in measuring CPU supply and demand is that such measurements must be performed on both the mobile computer and the remote server(s) used in cyber foraging. Further,

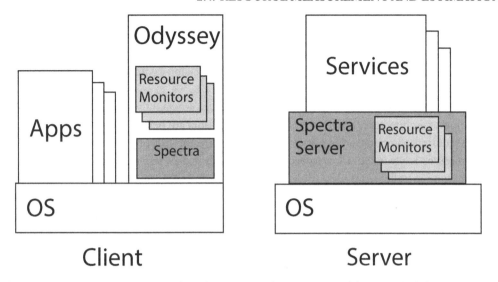

This figure shows the architecture of the Spectra cyber foraging system (shown as shaded components in the figure). Both the mobile computer and remote sever run resource monitors, each of which is responsible for measuring and estimating the supply and demand of a particular resource. The resource monitors on the remote server periodically and asynchronously report the state of the server by sending remote procedure calls to the mobile computer.

Figure 2.4: Spectra architecture.

the remote server employed for cyber foraging may change depending on such factors as the location of the mobile computer. This means that the cyber foraging system must either thoroughly measure behavior on all computers before making an accurate decision, or it must account for differences in processing power across computers. Spectra and Chroma take the latter approach by using the processor speed to scale expectations for computation time. However, processor speed is an imperfect predictor of computation time, as some application activities are memory or I/O bound.

A second challenge arises from the distributed nature of cyber foraging. The partitioning decision is made on one computer (often, but not always the mobile computer) using as an input the CPU load of other computers. Thus, when a partitioning decision arises, the cyber foraging system could potentially query all other computers that might participate in the computation for their current CPU load. Unfortunately, this involves a network round trip. If the computation being considered is relatively small, then the extra time to perform the query of remote state could add an unacceptable overhead relative to the compute time. (The opposite is also true: if the computation is long-running, then the relative overhead of a network round-trip may be insignificant.)

Due to this concern, systems like Spectra use slightly stale estimates of remote state. As shown in Figure 2.4, Spectra runs *resource monitors* on both the mobile computer and remote servers. Each monitor is responsible for a different resource such as CPU, network, etc. The resource monitors on the remote server periodically and asynchronously report the state of the server to the mobile computer via remote procedure calls. To avoid a network round trip, Spectra uses the last received estimate in its partitioning decision, even though this estimate may be several seconds old. Other systems, such as MAUI, make the entire partitioning decision asynchronously and thereby also avoid synchronous state updates.

Most current cyber foraging systems do not yet adequately account for two modern trends in computer architecture. First, variable-speed processors are pervasively used in mobile computers. If the required computation is not expected to require the full processing power of the computer, the processor is placed in a low-speed state in which it uses less energy to perform the computation. Second, parallelism is ubiquitous, especially in server design, although processors that employ multiple cores. For computation that can be parallelized, e.g., using multiple threads, the amount of available cores on a target computer is an important factor in determining how fast the computation will complete.

Odessa [Ra et al., 2011] is perhaps the first cyber foraging system that attempts to explicitly account for processor parallelism. It supports applications that utilize additional cores through data parallel computation. Odessa assumes that utilizing an additional core will provide an ideal (linear) speedup as long as spare cores are available to host the computation. After adjusting parallelism (for instance, by spawning a new worker thread), Odessa monitors application performance (makespan and throughput) to verify that the change in parallelism had the expected effect. This allows Odessa to self-regulate when applications hit a scalability wall.

In summary, with a few exceptions, current cyber foraging systems use relatively simple methods to estimate CPU supply and demand. Thus, there currently appears to be a substantial opportunity for further investigation to develop more sophisticated models, especially ones that account for variable-speed and multicore processors.

2.4.2   NETWORK

Cyber foraging systems measure network supply (i.e., the available bandwidth and latency) through a variety of strategies: active measurement, passive measurement, or a combination of the two. Passive measurements observe traffic sent over the network for other purposes; they do not inject additional traffic. This conserves both battery energy and cellular network usage. However, passive measurements may become stale if the mobile computer does not send or receive data for a time. Active measurements, which inject traffic into the network solely for the purpose of measuring the network performance, provide more recent data.

Spectra is an example of a system that uses passive measurements. Its network monitor predicts bandwidth and latency using an algorithm first developed for Odyssey [Noble et al., 1997]. Its RPC package logs the sizes and elapsed time of short exchanges and bulk transfers. The small RPCs are used to predict latency, and the larger RPCs are used to predict throughput. Spectra periodically examines the total bandwidth available to the mobile computer. It then estimates the bandwidth likely to be available for communicating with each server, assuming that the first, wireless hop in the network is the bottleneck link. Passive measurement of network demand (bytes transmitted) is accomplished in a similar manner; Spectra simply queries the RPC package to determine the bytes required to transfer state for each RPC performed.

MAUI is an example of a system that uses both active and passive measurements. MAUI employs the simple strategy of sending 10 KB of data to the remote server and measuring the throughput of the transfer directly to obtain an average throughput. The developers of MAUI found this works well because the 10 KB size is representative of the typical transfers to the MAUI server. MAUI also uses passive estimation to refine its active measurements. Every time that a method is offloaded, MAUI obtains a more recent observation of network quality by measuring the duration of the transfer. Active measurements are only employed if no transfers have provided recent passive estimates; if more than a minute passes without a transfer, MAUI sends 10 KB to the server and observes the result.

In order to avoid overreaction to transient changes in network quality, most cyber foraging systems use filters to smooth recent observation before using those observations to predict future network conditions. Spectra and Chroma adopt Odyssey's use of an exponentially weighted moving average. MAUI uses a sliding window. More sophisticated strategies are possible. For instance, one comprehensive study [Kim and Noble, 2001] surveyed a wide variety of strategies for combining recent observations to predict future network quality. The study found that a *flip-flop* filter worked best. This filter uses two exponentially weighted moving average filters. One is agile, with a gain of 0.1, and the other is stable with a gain of 0.9. A controller selects between the two, trying to employ the agile filter whenever feasible, but falling back to the stable filter with the observations are unusually noisy.

Just as with CPU measurement, current hardware trends are making network estimation more challenging. One such trend is the use of power-saving modes for network interfaces. Because the network interface is one of the dominant sources of energy consumption for small, mobile computers, many such computers place the wireless interface in a low-power mode when it is not sending or receiving packets. For instance, the 802.11 standard [IEEE Local and Metropolitan Area Network Standards Committee, 1997] defines a power-saving mode (PSM) that disables the interface for short periods (typically 100 ms). Since the interface cannot receive packets when it is disabled, the 802.11 base station queues data until the interface reactivates. Unfortunately, use of PSM can cause substantial performance degradation for short request-response traffic [Anand et al., 2003]. The mobile computer sends the request and waits for a response. The network interface enters power-saving mode shortly after the request is sent since it sees no network traffic. The response arrive a short time later but the base station must delay sending it for almost 100 ms until the network interface of the mobile computer reactivates. This can lead to more than an order of magnitude slowdown for short RPCs.

The authors of MAUI observed that this effect is a substantial problem for cyber foraging systems, especially those that try to offload work as small as a single method. Even for larger transfers, the degradation of disabling the interface can be seen due to TCP slow-start. Initially, the sender has a small window size and sends only a few packets. This causes the interface to go to sleep before the TCP acknowledgments arrive, leading to an unnecessary delay. Only when the network interface is continually busy (after slow-start ends) does the effect dissipate. The MAUI authors report that for short transfers (less than 75 ms round-trip time) disabling PSM both saves energy and reduces communication time. When round-trip time approaches 100 ms, enabling PSM saves energy. Unfortunately, some operating systems such as Windows Mobile do not currently expose interfaces that allow applications to enable or disable PSM. However, research proposals that allow applications to provide hints that guide operating system power management decisions [Anand et al., 2003] could potentially allow cyber foraging systems to adapt network power management to improve performance and save energy.

A second trend that complicates network measurements is that mobile computers often have more than one connectivity option. For instance, a cell phone might choose to send data over a WiFi network (if it is located near a hot spot) or a cellular network (if it has a data plan). Additional networks may have different characteristics (e.g., higher bandwidth, lower latency, or a lower loss rate), and the best network to use may change over time or as a user moves. Further, since cellular data plans are increasingly enforcing limits for the amount of data that may be sent within a time period, recent protocols for mobile computers have proposed opportunistic offload of communication from 3G cellular to WiFi networks [Balasubramanian et al., 2010]. Finally, the performance of offloaded computation might benefit from applying strategies that either stripe data over multiple networks or route traffic over the most appropriate of the available wireless network [Balasubramanian et al., 2010; Higgins et al., 2010; Schulman et al., 2010].

Thus, cyber foraging systems may need to simultaneously consider conditions for more than one wireless network. However, predicting the bandwidth and latency that can be achieved when using more than one network is more complicated than predicting those values for a single network.

2.4.3   BATTERY ENERGY

The supply of battery energy is one of the simplest resource values to measure and estimate, whereas the demand for battery energy is one of the most difficult. The supply of battery energy is just the amount of charge remaining in the battery. Most mobile computers provide standard interfaces for querying this value [Intel and Microsoft, 1996; Intel et al., 1998].

Energy demand is the amount of batter energy consumed by a particular computation. Cyber foraging systems typically use one of two methods to estimate this value. The first method is direct measurement. Spectra and Chroma both use this approach. They measure the amount of energy in the mobile computer's battery before an operation begins and after the operation completes. The difference between the two values is the amount of energy required to perform the computation.

Direct measurement has two drawbacks. First, if more than one activity occurs simultaneously, it is difficult to separate the energy cost of each activity. Spectra adjusts for this issue by discarding all energy demand measurements in which operations executed concurrently. A second drawback is that many mobile computers do not provide fine-grained measurements of battery capacity. Spectra used a Smart Battery [Dallas Semiconductor Corporation, 1999], a gas gauge chip that reports detailed information about battery state and power usage. However, many mobile computers lack equivalent chips and report battery energy at granularities as coarse as 1% of total battery capacity — such a granularity is insufficient to measure brief computational events.

Consequently, many cyber foraging systems use a model-based approach to estimate energy demand. This approach executes a series of microbenchmarks on the mobile computer in a laboratory setting while an external power measurement device measures how much energy the computer expends executing those benchmarks. An energy model is constructed using the results of these experiments. The model assigns specific energy costs to measurable activities performed by the mobile computer such as executing a given amount of computation, sending or receiving a byte of data over a wireless network, or enabling the screen backlight. Such models are naturally specific to a particular make and model of a mobile device. Given such an energy model, a cyber foraging system can measure the occurrence of events such as computation and network transmission that occur during application execution, then use the model to assign an energy costs to those events. The end result is an estimate of the energy used to perform the activity.

The Ecosystem project [Zeng et al., 2002] developed some of the first energy models for use by mobile computing software. Early mobile computers such as the Compaq Itsy often provided detailed specifications [Flinn et al., 2000] from which such models were developed. Since then, the model-based approach to energy measurement has proven popular, and many new models have been developed. For example, the PowerTutor project [Zhang et al., 2010] recently created energy models for several smart phones. Many cyber foraging systems, including both MAUI and CloneCloud, employ energy models to estimate power demand.

A drawback of using energy models is that they necessarily do not capture all the potential factors that may influence mobile computer energy usage. As an example, a model might capture the average energy consumed per CPU cycle of computation. However, different instructions consume different amounts of energy, so the mix of instructions can affect the total energy consumed by an operation. Further, the amount of memory accesses per instruction substantially changes the energy expenditure [Martin, 1999]. Finally, processor power management will substantially alter energy usage. Thus, in developing any energy model, there is a trade-off between complexity and accuracy. It is possible to capture any one of the above effects given sufficient microbenchmarks, offline analysis, and online accounting. However, capturing all possible effects for CPU, network, storage, display, and other components is a very daunting task. Most cyber foraging systems compromise by using simpler models and accepting that demand measurements will include a moderate amount of noise.

2.4.4   FILE CACHE STATE

Not all state resides in the application memory; some state exists in the file system or other persistent storage systems such as databases. One solution for partitioning applications that interact with persistent state is to execute the components that perform such interactions only on the computer that stores the state (e.g., only on the mobile computer). CloneCloud takes this approach.

A second approach is to use a storage system that replicates its contents across multiple computers. When the application reads or writes persistent data, the storage system ensures that modifications made on one computer are transferred to other computers so that each component sees the latest version of data no matter where it executes. Spectra is an example of a cyber foraging system that takes this approach; it uses the Coda distributed file system [Kistler and Satyanarayanan, 1992] to store persistent data.

However, replicated storage systems such as Coda work by caching file content on each computer. Data accesses can take significant time and energy when items are unavailable locally and the items need to be fetched from another location. Systems such as Spectra therefore monitor file cache state as a resource to estimate replication costs when making partitioning decisions.

Spectra's file cache state monitor interacts with Coda to predict cache state. It asks Coda which files are cached locally on the computer. Although it is possible that cache state will change slightly during a computation, those changes are unlikely to be significant. The Spectra monitor also obtains from Coda an estimate of the rate at which uncached data will be fetched from a file server.

The state of the cache is the supply measurement for this resource. The demand measurement is the set of files accessed by an application component. Spectra measures this passively by interposing on file system accesses as application components execute.

While Spectra and Chroma are currently the only cyber foraging systems to consider file cache state as a resource, the importance of this resource may be increasing over time. A recent study [Kim et al., 2011] observed that storage is a surprising bottleneck for many applications such as Web browsers running on mobile phones due to the poor interaction of flash storage with middleware layers that are overly conservative in performing synchronous writes. It may therefore be necessary to consider file I/O more carefully when estimating the performance and energy costs of candidate partitions.

2.4.5   MEMORY

Chroma is the only cyber foraging system to measure memory. It uses two different algorithms to discover the supply of available memory on a computer [Balan, 2006]. On Linux 2.2 kernels, Chroma reads the available free memory and the space available in raw disk blocks used for relatively temporary storage from the /proc file system. It then estimates memory supply by adding these two values together and subtracting a small amount of memory (4 MB) to ensure that there is always some spare capacity available.

On Linux 2.4 kernels, Chroma estimates supply by adding the amount of free memory and the amount of memory in clean, inactive pages (because such pages are easy to convert to free pages). Chroma does not include dirty or active pages in the amount of free memory since reclaiming those pages requires substantial effort from the operating system — thus, those pages can not be used relatively quickly.

2.5   DEALING WITH ERRORS DURING PARTITIONING

The previous sections in this chapter describe the partitioning decision process implemented by current cyber foraging systems. The discussion so far has not considered the impact of errors that can arise during partitioning such as inaccurate predictions and unexpected changes in the mobile environment. The reason for this omission is that most cyber foraging systems developed to date do not explicitly consider such errors or have only rudimentary strategies to cope with their occurrence. Therefore, in this section, we elaborate on a some of the sources of inaccuracy and failure during partitioning, and we speculate about some possible strategies that could cope with these errors.

As with any distributed system, one or more of the components of a cyber foraging system may fail during operation. The objective of most cyber foraging systems is to provide behavior equivalent to the behavior that would be seen if the application ran entirely on a mobile computer, albeit with better performance. If the mobile computer itself were to fail (for example, after running out of battery power), an application running on that computer would lose all non-persistent state. It is usually trivial for a cyber foraging system to emulate this behavior by canceling any remote operations when a mobile computer fails.

Failure of a remote component such as a wireless network or a remote server requires more effort to mask. If the cyber foraging system has offloaded computation to a remote computer that fails, the state of the computation is lost. Even if the failure is due to a network partition, waiting for a network connection to be reestablished may introduce unacceptable delay for interactive applications.

The cyber foraging systems that explicitly consider such failures often use simple timeout mechanisms to detect and recover from them. For example, MAUI uses a timeout to detect that it has lost contact with the remote server while that server is executing a remote application component. At that point, it reconsiders the partitioning decision and re-executes the component either locally or on a different server. This strategy assumes that the state required to restart the computation (i.e., a checkpoint) still exists — this assumption holds for most cyber foraging systems because the application blocks waiting for the remote operation to complete. Most cyber foraging systems also forbid remotely executed components from externalizing state (e.g., sending output to a screen, network, or other device) in order to avoid the complexities that arise from distributed checkpoint and restart [Elnozahy et al., 2002].

The difficulty in employing timeout-based strategies lies in choosing the correct timeout value. If the timeout is too short, the cyber foraging system will abandon some remote computations that would shortly complete after a transient failure. Such transient failures are especially common in mobile computing due to the vagaries of wireless networking. On the other hand, if the timeout is set too long, then the cyber foraging system will wait unnecessarily for a remote operation to complete before initiating a recovery action. This will increase interactive delay. There is no perfect timeout value: some failures will be permanent, some transient failures will require an excessively long time to recover, any it may be possible to recover from other transient failures almost immediately. Even for those few cyber foraging systems such as MAUI that employ some timeout-based strategy, there has been no research on how to best set such timeouts.

Changes in the mobile environment can produce concerns that are related to those introduced by the possibility of remote component failure. Even if the cyber foraging system can accurately predict the current resource availability at the time a component begins executing, those conditions can change dramatically for the worse during the execution of the application component. For instance, the load on a shared remote server may increase substantially, or the bandwidth available over a wireless network may decrease rapidly. The same timeout-based strategies employed to survive remote component failure can also address such scenarios. The timeout for a remote operation will expire because the operation cannot complete on time due to increased server load or reduced network quality, and the mobile computer will restart the computation after reevaluating the partitioning decision.

Unlike explicit failures, more information may be available when resource availability unexpectedly fluctuates. A cyber foraging system could potentially re-evaluate the partitioning decision based on new predictions of future availability and choose to either continue the computation as it is currently executing or restart it using a new partitioning strategy. Such a decision should also include some estimate of the progress that has already been made, either based on the measured resources available during the execution so far or based on an explicit measurement of application progress. Further, a cyber foraging system might also consider changing the partitioning strategy if resource availability *improves*. For instance, it could abandon a computation currently executing on the mobile computer and re-execute it on a newly discovered remote server. No cyber foraging system to date currently re-evaluates its partitioning decision in this manner, perhaps due to the added complexity.

In addition to resource supply fluctuating during a computation, an application's demand for resources may differ substantially from what was initially predicted. For instance, how should a cyber foraging system react if an application has already used twice the amount of predicted CPU cycles for a computation? Even if the cyber foraging system were to re-evaluate its partitioning decision, it needs a new estimate of demand since the prior estimate was clearly inaccurate. One solution may be to predict resource demand (and potentially resource supply) as probability distributions, rather than as specific values. Given such a distribution, the cyber foraging system could calculate the conditional expectation for additional CPU cycle consumption given that the computation has already used a certain amount of cycles. Based on this revised estimate, the cyber foraging system might choose to change the partitioning strategy.

Redundancy is a potential solution to the above concerns that has not yet been fully explored in the context of cyber foraging. No system explicitly considers the possibility of running the same application component simultaneously on the mobile computer and on a remote server.

Chroma [Balan et al., 2003] and Slingshot [Su and Flinn, 2005a] both have the ability to execute the same application component on more than one server at the same time. Chroma allows its users to specify *over provisioned* tactics that utilize extra servers to improve execution time. A specified tactic may run the same components on more than one server and use the fastest response to hedge against load spikes. When allowed by application semantics, Chroma can also decompose a component and parallelize it across multiple servers, or it can run components at different fidelities on different servers and use the highest fidelity response returned within a specified amount of time.

Slingshot always runs remote application components on a well-known home server, but its users can explicitly designate additional nearby servers on which to run the same components. This is similar to the strategy used in Chroma to guard against load spikes, except that Slingshot expects the nearby server to have a faster response time in the normal case because of its proximity to the mobile computer.

While both Chroma and Slingshot provide mechanisms for execution of application components on multiple servers, neither provides a policy for automatically deciding when to employ such redundancy. Both Chroma and Slingshot currently leave the decision to employ redundancy up to their users. Thus, future research could explore methods for deciding when to employ redundancy in such cyber foraging systems.

Redundant execution is a standard fault-tolerance technique for masking the failure of one or more computational nodes [Schneider, 1990]. Given that failures are especially common in mobile computing due to the use of unreliable wireless networks and battery-powered components, it seems reasonable to consider replicating important computations in more than one location to tolerate a network disconnection or node failure with minimal performance impact. Such replication may be active, in which case redundant replicas are started concurrently, or passive, in which case an additional replica is started on demand. Passive replication seems an attractive replacement for timeout-based cancel-and-restart of an unresponsive computation. If the cyber foraging system loses contact with a remote server, it could start an additional computation locally or on another server, use the first response it receives (whether it comes from re-establishing contact with the lost server or from completing the newly started computation), and canceling the computation that has not yet responded.

Systems that replicate computation must carefully handle operations that require exactly once semantics, such as sending network messages to external parties. During failure-free operation, one replica is typically designated the primary and is responsible for sending such messages. Messages from other replicas are suppressed. When a failure occurs, message logging techniques can ensure that only one instance of an exactly once operation is performed. Alternatively, protocols can be designed so that duplicate messages sent during failover are dropped (for example, a sequence number can be added to all messages).

Redundant execution is also a potential solution to dealing with uncertainty in predicting resource supply and demand. Given such uncertainty, running a component in more than one location will decrease the expected time to completion since the cyber foraging system can simply use the results of the first replica to complete. If resource estimates are expressed as probability distributions, it is possible to calculate precisely the expected improvement in execution time. However, redundant execution may often consume more energy and may use more of other limited resources such as cellular data. Thus, it may be best to simply consider redundant plans as additional options in the partitioning decision, apply the same procedure used to balance competing concerns and choose from among both redundant and non-redundant strategies.

In summary, current cyber foraging systems rarely consider the complexities that arise from failures, changes in the mobile environment, and imprecise predictions of resource supply and demand. Most cyber foraging systems to date have considered offloading only relatively brief operations such as recognizing a spoken utterance or finding faces in an image. As mobile applications grow in complexity, the scope of offloaded components may increase — in such cases, the penalty of component failure or mispredictions will be larger. Therefore, the importance of coping with such errors may increase. Fortunately, there exists substantial opportunity to improve the state of the art by incorporating fault-tolerant design, predictions that quantify uncertainty through probability distributions, and strategic use of redundancy.

2.6 APPLICATIONS

The cyber foraging systems discussed in this chapter support a wide variety of applications. We list these systems in order from the oldest to the newest. RPF executes compile tasks. Spectra runs a speech recognizer, document preparation tools, and a natural language translator. Chroma supports face recognition, text-to-speech synthesis, image filtering, 3D model viewing, speech recognition, music identification, natural language translation, lighting for 3D models, and creation of 3D scenes from 2D images. MAUI executes a face recognizer, a video game, and a chess game. CloneCloud runs a virus scanner, an image search application, and privacy-preserving targeted advertising. MARS targets three computationally intensive applications: augmented reality pattern recognition, face recognition, and an augmented reality video game. Odessa supports face recognition, object and pose recognition, and gesture recognition.

Several conclusions can be drawn from examining this list of applications. First, the tasks best suited for cyber foraging are compute-intensive. Such tasks benefit most from remote execution, and they therefore offer the best opportunity to recoup the costs of sending state over the wireless network.

Second, cyber foraging is not necessarily limited to batch tasks like compiling and document processing. Most of the applications listed above are interactive. However, all have significant compute chunks that occur between interactions with the user. The length of such compute chunks must be substantial compared to the latency of the wireless network, unless the computation can be performed asynchronously.

Third, most of the recent applications that leverage cyber foraging involve multimedia, either through intense audio processing (e.g., speech recognition and natural language translation), image processing (e.g., face recognition), or support for video (e.g., games). Thus, the trend to incorporate more multimedia in everyday applications may increase the opportunities to employ cyber foraging in future applications.

Finally, many of the applications listed above are especially relevant to mobile computer users. Applications such as natural language translation have clear applications for visitors traveling through a foreign country. Face recognition and music identification are already used in popular mobile applications. Thus, the future seems promising for cyber foraging: such applications demand capabilities that mobile computers may simply be unable to provide with purely local resources while running with a limited battery capacity.

2.7   SUMMARY

This chapter has pursued a functional decomposition of the partitioning decision. We first examined the different goals that cyber foraging systems aim to optimize when partitioning applications. Figure 2.1 summarized the goals pursued by several different systems discussed in this chapter. We next discussed how these systems select candidate partitions — Figure 2.2 summarized the different approaches. We then looked at how the various systems choose one of the candidates. In order to gather the necessary information to make this choice, most systems monitor the supply and demand of one or more resources. Figure 2.3 captured the different approaches used.

We next described some opportunities for future research in this domain. In particular, we noted that most cyber foraging systems do not sufficiently account for the possibility of component failure, changes in environmental conditions, and inaccuracy in resource predictions. We noted that incorporating predictors that explicitly include measurements of uncertainty and using replication are two solutions that could potentially help address these problems.

Finally, we examined the applications currently supported by cyber foraging systems. We noted that most such applications are compute-intensive. When the applications are interactive, there are usually significant compute chunks between user interactions. The most recent cyber foraging systems have applications that incorporate multimedia and which are particularly relevant to mobile users. We noted that these characteristics match will with current trends in mobile applications, indicating that the benefit of cyber foraging may continue to increase in the future.

CHAPTER 3

Management

In this chapter, we discuss issues related to the location and management of servers that host remote operations during cyber foraging. Following the terminology introduced in the early literature [Satyanarayanan, 2001], we refer to a remote computer that may temporarily assist a mobile computer as a *surrogate*.

The main questions discussed in this chapter are the following.

1. Where should surrogates be located?

2. How should remote operations on a surrogate be isolated?

3. How should surrogates acquire and manage the state associated with cyber foraging?

4. What are the steps required to discover and start using a surrogate?

Section 3.1 discusses several factors that impact the location decision, including network latency and bandwidth, ease-of-management, cost, and privacy concerns. It then describes several options for location: deploying surrogates as close to the edge of the network as possible, use of dedicated computers in a data center, and opportunistic use of computers in a data center. It describes how any of these options may be best, depending on which of the factors listed above are considered the most important.

Section 3.2 discusses how remote operations on a surrogate may be isolated. Early cyber foraging systems relied on the process-level isolation provided by the operating system. More modern cyber foraging systems have leveraged virtualization, either through emulation of a specific hardware architecture (such as the x86 ISA) or through application virtualization provided by managed run-times for languages such as Java and C#. The abstraction chosen for isolation also affects programmer flexibility and transparency. Java Virtual Machines and other managed run-times are designed to be portable across many different types of computer hardware. In contrast, process-level operation often restricts the operating system type and version that a surrogate must run. The isolation choice also impacts the amount of state associated with an operation, which may in turn affect performance overhead and start-up costs.

Section 3.3 describes issues that arise from caching data and computational state on surrogates. Cyber foraging systems differ greatly in the persistence of data stored on surrogates. In some systems, data persists only for the lifetime of a single operation. Alternatively, data may persist between remote operations or even remain cached after a mobile computer disconnects from a surrogate. This section also discusses several methods for acquiring the application-specific state needed to perform operations.

Section 3.4 describes the mechanics whereby a mobile computer can discover and use a surrogate. It also discusses management issues that arise from hosting multiple operations on a surrogate such as admission control.

3.1   SURROGATE LOCATION

When deploying a cyber foraging system, one must decide where to locate the surrogate computers that host remote operations. It is most useful to characterize the location decision according to the network proximity between the mobile computer and the surrogate.

3.1.1   SURROGATES NEAR MOBILE COMPUTERS

The original case for cyber foraging [Satyanarayanan, 2001] envisioned that surrogates would be placed as near to mobile computers as possible: "As computing becomes cheaper and more plentiful, it makes economic sense to 'waste' computing resources to improve user experience. Desktop computers at discount stores already sell today for a few hundred dollars, with prices continuing to drop. In the foreseeable future, we envision public spaces such as airport lounges and coffee shops being equipped with compute servers or data staging servers for the benefit of customers, much as table lamps are today." In this vision, surrogates are as close as a single WiFi network hop from the mobile device.

At the other extreme, most current distributed mobile applications connect cell phones and tablets to servers in the cloud. Such servers may reside in a managed data center that is very distant in terms of network topology — often tens of milliseconds away, and possibly hundreds of milliseconds if the data center is located on another continent or communication must traverse a cellular network. Even for conventional applications, this amount of network latency degrades the user experience, leading cloud developers to replicate services across geographically distributed data centers [DeCandia et al., 2007] or cache content closer to clients [Nygren et al., 2010].

Locating surrogates closer to mobile computers has substantial performance benefits, especially for highly interactive applications. As discussed in the previous chapter, if a synchronous operation is offloaded to a surrogate, the operation will complete faster only if the reduction in computation time is greater than the time required to transfer data and control over the network. As the distance (in terms of network latency) between the mobile computer and the surrogate increases, larger and larger computation chunks are required to recover the time lost due to network communication. Network latency has a similar effect on energy usage because the mobile computer expends power while waiting for the surrogate to return a result. In the limit, it may not be feasible to offload any part of an application if a surrogate is located too far away.

The authors of MAUI characterized the effect of network latency on their system [Cuervo et al., 2010]. For one application (a chess game) offloading methods to a surrogate improves performance only if the surrogate is located nearby — with a network round-trip time (RTT) of 25 ms or less. Offloading methods to a more distant surrogate hurts performance, and so MAUI does not benefit from cyber foraging in those circumstances. For other applications (face recognition and a video game), the performance difference between using a nearby (10 ms RTT) and a distant (220 ms RTT) surrogate can be as large as 50%. When the MAUI authors considered energy usage, they found that two applications (chess and the video game) consumed more energy when the mobile computer offloaded methods over 3G (incurring a 220 ms RTT) than if the method had not been offloaded, whereas WiFi transmission, which led to lower round-trip times, enabled offloading to always decrease the energy usage of a cell phone.

3.1.2   SURROGATES IN THE CLOUD

Locating surrogates in data centers does have several benefits. One such benefit is ease of management. It is much easier to replace a hardware component that has failed in a centralized data center than it is to make a service call to a remote location to replace the same component. Physical proximity may make it easier to troubleshoot software problems and repair misconfigurations. It is likely that network connectivity, power, and other external inputs to the surrogate are more reliable in a data center than in a coffee shop or airport lounge.

Locating surrogates in a data center also makes it easier to provide physical security. If surrogates are deployed in public locations, then additional measures, discussed in the next chapter, may be required to prevent tampering.

Finally, locating surrogates in cloud data centers can improve the performance of offloaded computation when that computation interacts with other cloud components or accesses data stored in the cloud. If such a computation is hosted near the mobile computer, then each communication with cloud resources may incur high latency and/or limited throughput, especially if the last-hop link connecting the surrogate to the cloud is poor. This property can be viewed through the lens of a partitioning decision. If an application component interacts with data and other applications hosted in a data center more than it interacts with the mobile computer, then it will usually make more sense to execute that component on a surrogate located in that data center rather than on a surrogate located near the mobile computer.

3.1.3   SURROGATE LOCATION FOR CURRENT CYBER FORAGING SYSTEMS

For these and other reasons, the cyber foraging systems developed to date have chosen different options for the location of surrogates. We next describe these choices in more detail.

Many systems target surrogates located as close to mobile computers as possible. Following the popular terminology introduced recently [Satyanarayanan et al., 2009], we will refer to such deployments as "cloudlets." A cloudlet is defined to be "a trusted, resource-rich computer or cluster of computers that's well-connected to the Internet and available for use by nearby mobile devices." While the cloudlet vision included many diverse components, the essential feature of a cloudlet is that it can provide low-latency, one-hop, high-bandwidth wireless connectivity to mobile devices through physical proximity. This vision requires that cloudlets be physically dispersed at a scale much greater than that used by current Internet service deployments. Much like current WiFi access points, cloudlets might be deployed in coffee shops, doctors' offices, and other places of business to provide reasonable coverage for mobile users. In fact, the network requirements for cloudlets make the physical pairing of public WiFi access points and cloudlets especially attractive.

As previously mentioned, the notion of locating surrogates as close as possible to mobile computers was proposed as part of the original vision for cyber foraging [Satyanarayanan, 2001]. Systems building on this vision such as Spectra and Chroma have adopted the same deployment model as part of their designs.

One alternative to cloudlets is to place surrogates in managed data centers. Systems such as MARS [Cidon et al., 2011] and CloneCloud [Chun et al., 2011] have adopted this approach. Other cyber foraging systems such as Maui do not explicitly choose between the cloudlet and cloud models. Instead, they evaluate their systems across both scenarios.

Cloudlet deployments are inherently shared among multiple users. The scenarios most often suggested for cloudlet deployment are public locations with many mobile computers. In addition, the need for physical proximity requires that mobile computers connect to different cloudlets as their user moves from location to location. Reserving dedicated resources at every location that a user might possibly visit is wasteful and likely infeasible.

Surrogates deployed in the cloud, however, may be either dedicated or shared. Since the location of surrogates within the cloud does not impact the quality of the network connection in the same manner that the location of surrogates outside the cloud does, it is reasonable for a given user to use the same cloud surrogate repeatedly. Thus, reserving dedicated resources for each user on such surrogates is not particularly wasteful. In addition, data center consolidation techniques such as those used in Amazon's EC2 can multiplex multiple virtual surrogates on the same physical host. This provides the illusion of dedicated resources, when in fact such resource are really being allocated on demand.

Slingshot [Su and Flinn, 2005a] is an example of a cyber foraging system that leverages a dedicated cloud resource. Slingshot assumes that there exists a surrogate, called the *home server*, that is dedicated specifically to each mobile user. It executes remote operations on a user's dedicated surrogate to ensure reliability and correctness. At the time Slingshot was proposed, the authors envisioned using a remote server under a user's control as a surrogate. However, the recent success of EC2 and other cloud hosting services make such platforms a more attractive choice for hosting dedicated surrogates today.

Slingshot combines multiple deployment models to try and achieve the best of both worlds. In addition to offloading operations to a dedicated cloud server, Slingshot also opportunistically executes additional instances of remote operations on cloudlets. A Slingshot client broadcasts requests to perform remote operations to both cloud and cloudlet surrogates. The client uses the first response it receives — typically, this response will be from the cloudlet surrogate because the physical network proximity provides better performance. The cloud surrogate provides fault-tolerance, however. If the cloudlet fails, or if it provides poor performance due to a spike in demand for the shared resource or the mobile user moving away from the surrogate, then the cloud surrogate will provide the faster (or only) response. The cloud surrogate provides additional benefits that will be explored in subsequent sections: it verifies the correctness of operations performed by cloudlet surrogates, provides reliability for persistent services, and helps hide the performance cost of initializing cloudlet surrogates.

3.2   ISOLATION OF REMOTE OPERATIONS

The choice of how to isolate operations running on a surrogate is orthogonal to the choice of where to locate surrogates discussed in the previous section.

The need for isolation of surrogate operations is fundamental. Almost all cyber foraging systems uphold the following *result-equivalence* property: *The observable results of an operation that executes remotely on a surrogate should be indistinguishable from results that could have been produced by the same operation if it had executed on the mobile computer.* The result-equivalence property enables the transparent migration of functionality to surrogates as conditions allow; the cyber foraging system does not need to interact with the user to decide whether such migration is correct because the result-equivalence property ensures that the results of a remote operation can always be correctly used in lieu of performing the operation locally.

In order to guarantee result-equivalence, a cyber foraging system must be able to characterize the possible results of remote operations. This is typically accomplished by *isolating* the operation within an execution sandbox on the surrogate. The execution sandbox has several functions. First, it provides a set of dependencies such as libraries, operating systems, and services — typically, these dependencies are the same as those on the mobile computer. Thus, an operation that invokes a dependency on the surrogate can expect a result that could have been returned by invoking the same dependency on the mobile computer. Second, the execution sandbox typically limits and/or mediates the interactions with dependencies external to the sandbox. For instance, a sandbox might prevent an operation hosted on a surrogate from invoking services provided by the surrogate's operating system because that operating system may not be the same as the operating system running on the mobile computer (and hence may return results impossible to generate on the mobile computer). Finally, the execution sandbox may suppress duplicate messages and other external effects for cyber foraging systems that replicate application components on multiple surrogates, as discussed in Section 2.5.

There are many possible methods for providing execution sandboxes. We refer to this choice as the *unit of isolation* for a surrogate — the decision is one of the more substantive design choices in deploying surrogates for a cyber foraging system.

3.2.1   DESIGN GOALS

There are several goals to consider when choosing the unit of isolation. The first goal is *safety*, or the degree to which result-equivalence is guaranteed. Ideally, the isolation mechanism should prohibit all operations that would enable a remote operation to produce a result that it could not have produced via a local execution. As will be described below, some isolation mechanisms do not provide full result equivalence; they must rely on the programmer who implements the remote operations to provide code that behaves in certain ways. Other systems guarantee result equivalence for all code that runs remotely, but they may restrict some functionality to execute only on the mobile computer.

Thus, the second goal is *flexibility*; it is desirable to support remote execution of the greatest number of possible operations. Clearly, as more operations are supported, the opportunities for an application to benefit from cyber foraging increase because more components become eligible for offloading. Further, as the number of operations that *cannot* be offloaded decreases, the size of the application execution blocks between components that must be executed locally will increase. Since the benefit of cyber foraging is usually greater for larger execution chunks, increasing the chunk size will often create new opportunities to benefit from cyber foraging where no such opportunity existed before (because all execution chunks were too small).

The third goal when choosing the unit of isolation is *transparency*. Ideally, the execution environment on the surrogate and on the mobile computer would be the same, so the same binaries can be used to execute an operation in either location. Such a system is fully transparent. Alternatively, it might be feasible to dynamically transform binaries compiled for the mobile computer to generate equivalent executable code for the isolation unit on the surrogate. Relaxing the goal of transparency further, a developer might perform such translation statically, for instance, by compiling two different versions of a program. A still less transparent solution might require programmer assistance to invoke run-time services, libraries, or middleware components that would then execute code either locally or remotely.

The fourth goal is to minimize the performance *overhead* of isolation. Some sandboxing techniques such as hardware virtual-machine monitors (VMMs), add substantial memory, disk, and CPU requirements to those of the hosted operation. Techniques such as para-virtualization [Whitaker et al., 2002] can substantially reduce the resource requirements. The choice of isolation technique therefore directly impacts surrogate scalability: surrogates with lightweight isolation sandboxes can support more concurrent operations than those with heavyweight isolation.

The final goal, *fast start-up*, is strongly correlated with the fourth goal of reducing overhead. Heavyweight isolation mechanisms such as hardware VMMs can require a considerable amount of time to initialize the isolation sandbox and begin executing remote operations on behalf of a mobile computer. If the mobile computer has a lengthy interaction with a surrogate, then such start-up costs may be amortized across many remote operations. However, if a mobile computer has only a short-lived interaction with a surrogate, then it may not be feasible to use an isolation mechanism that requires a long start-up time.

3.2.2   PROCESS-LEVEL ISOLATION

Since the above goals often conflict with one another, cyber foraging systems have chosen many different isolation methods to sandbox operations on surrogates. Perhaps due to the scarcity of robust virtualization methods at the time, the earliest cyber foraging systems mostly used the process-level isolation mechanisms provided by the operating system. For example, RPF, Spectra, and Chroma all execute each remote operations on a surrogate with a separate process.

Spectra's isolation mechanism is typical of systems developed at this time. Application code components executed on Spectra surrogates are referred to as *services*. Each service executes as a

```
service_init (&argc, &argv);
while (1) {
    service_getop (&optype, &opid, path,
                       &indata, &inlen);

    rc = do_operation (indata, inlen,
                          &outdata, &outlen);

    service_retop (opid, 0, outdata, outlen);
}
```

This figure shows a portion of a program that executes a Spectra service. The program has an event loop that receives requests, executes the requested computation, and returns the result of the computation.

Figure 3.1: Sample Spectra service implementation.

separate process. Operating system isolation protects the surrogate server and other services from a malicious or faulty service. A service may either be started when the Spectra server begins executing or when the first RPC for that service is received. These two start-up methods balance the latter two goals for isolation: starting the service on server initialization provides better performance for the first request, but it may waste server resources as compared to running services only on-demand.

The Spectra server accepts multiple concurrent requests. Each RPC is handled by a separate thread, which routes the request to the appropriate service. Individual services may also handle multiple concurrent requests. When a server thread receives a RPC for a service, it assigns a unique identifier to the request and writes the request to the service's input pipe. When the service replies, it specifies the identifier of the request to which it is replying.

Spectra provides an application library that simplifies service implementation. Figure 3.1 shows the main loop of a simple service. The `service_init` function parses the command line and extracts Spectra-specific information. In the main loop, `service_getop` blocks until a request is received. The function returns the type of operation requested, a unique identifier associated with the request, and application-specific input data. The sample service has only one request type—services that handle more than one type of request multiplex on `optype`. When processing is complete, the service calls `service_retop`, passing in the request identifier and application-specific output data.

The process-level isolation used by Spectra and similar systems has low overhead, especially if services are only started on-demand, which prevents idle services from consuming resources. Since creating a new process is usually quite quick compared to network latencies between mobile computers and surrogates, the start-up delay of either method described above is also reasonable. Process-level isolation is also flexible since any code written for the mobile computer can execute on the surrogate.

Unfortunately, process-level isolation has a substantial downside: it provides poor safety. Spectra relies on the programmer to not invoke any operating system or other external service that would produce results on the surrogate that could not be produced on the mobile computer. A mistaken or malicious service provider can therefore cause considerable harm. Process-level isolation also provides poor transparency. Because mobile computers and surrogates typically have different hardware architectures (for example an ARM phone and an x86 server), separate code versions must be supplied for each new execution environment. This requires at a minimum static translation (i.e., re-compilation). Another substantial issue is differences in dependencies in the operating system and libraries. Unless such dependencies can be standardized across surrogates and mobile computers, it is likely that discrepancies will arise when an operation invokes all but the most basic of services provided by such dependencies. Lacking such standardization, cyber foraging systems that use process-level isolation will be more difficult to maintain than systems that use virtual machine isolation in which the virtual machine abstraction provides the desired standardization. The safety and transparency limitations of process-level isolation are so great that most recent cyber foraging systems have abandoned that isolation technique in favor of using hardware and application virtualization.

3.2.3   HARDWARE VIRTUALIZATION

Many modern cyber foraging systems have opted for the strong isolation of hardware virtual machines, each of which provides the abstraction of a virtual computer dedicated to running the software it encapsulates. Each virtual machine contains not just application software, but also the operating system, dynamic libraries, tools, and the rest of the software environment on which that application depends. As each virtual machine is self-contained, the software running in the virtual machine does not depend on the software environment of the host computer. This makes it feasible for surrogates to host heterogeneous virtual machines, which may have different versions of software installed or even support different operating systems. Because the hardware virtual machine virtualizes the processor interface, it is even possible to use dynamic translation to execute software binaries targeted for a different processor architecture than that used by the mobile computer processor. For instance, it is feasible to run ARM binaries on x86 processors via emulators such as QEMU [Bellard, 2005].

Slingshot is an example of a cyber foraging system that uses hardware virtual machine monitors: it runs the VMware VMM on each surrogate. The Slingshot paper [Su and Flinn, 2005a] cites two reasons for this choice. First, use of a hardware VMM simplifies the surrogate computing base, which consists only of the host operating system, the virtual machine monitor, and some surrogate management utilities. No configuration or setup is needed to enable a surrogate to run new applications because each VM is self-contained. Second, use of a hardware VMM provides both flexibility and transparency: Slingshot is shown to run applications without modifying source code even though the guest operating system required by those applications (Windows XP) differs substantially from the operating system running on surrogates (Linux).

Slingshot isolates each instance (replica) of an application inside a VMware virtual machine. Replica state consists of the *persistent state*, or disk image of the virtual machine, and the *volatile state*, which includes its memory image and registers. The persistent state is typically orders of magnitude larger than the volatile state. Slingshot loads persistent state on demand. It uses the Fauxide and Vulpes modules developed by Intel Research's Internet Suspend/Resume (ISR) project [Kozuch and Satyanarayanan, 2002] to intercept VMware disk I/O requests. Each block is fetched from a remote location and cached on the surrogate. The first access to a block may incur a delay as the data is transferred from remote storage, but subsequent accesses have low latency because they hit in the cache. Since the volatile state is smaller, Slingshot retrieves the entire volatile state from a remote location and loads it when the virtual machine is started.

The cloudlet vision includes the use of hardware virtual machine monitors for reasons similar to those cited by Slingshot: strong isolation makes surrogates easy to manage and deploy, and hardware VMMs provide flexibility by placing as few restrictions as possible on the software that can run on cloudlets. The cloudlet design emphasizes transient customization: pre-use customization and post-use clean up ensure that cloudlet infrastructure is restored to its pristine software state after each use, without manual intervention [Satyanarayanan et al., 2009].

The downside of hardware virtual machines is that they are heavyweight. Compared to other encapsulation methods such as process-level isolation, running software in a virtual machine consumes more CPU cycles, memory, and energy. The larger overhead of this method may therefore limit surrogate scalability. Start-up cost is perhaps an even more substantial concern. Because each VMM contains the entire state of a virtualized computer, it may take a considerable of time to transfer the state to a surrogate and install that state so that the application inside the VM can start handling requests from a mobile computer. Techniques such as Slingshot's loading of persistent state on demand can help reduce this start-up cost, but it is difficult to eliminate start-up latency entirely. The next section describes additional techniques for reducing start-up delay when using a new surrogate.

The overhead of hardware virtualization can be reduced through a technique known as *para-virtualization* that relaxes the isolation between the host operating system that runs on the surrogate and the guest operating systems running inside each VM. Goyal and Carter describe a cyber foraging system [Goyal and Carter, 2004] that uses Xen [Barham et al., 2003] for virtual machine encapsulation. Like heavyweight hardware virtualization, para-virtualization provides safety: applications running in different VMs cannot directly interfere with one another or access resources reserved for the surrogate machine. However, para-virtualization does not provide a clean abstraction of a virtualized computer for each VM; instead, the guest operating system inside each VM must be modified to interact with the VMM rather than access certain architectural features directly. This restricts the choice of software that can run on the surrogate: for instance, Goyal and Carter report that Xen at the time of their work supported only the Linux and BSD operating systems.

Goyal and Carter's system also supports process-level isolation via the Linux-Vserver, which encapsulates groups of processes using native Linux support such as the *chroot* system call. Goyal and Carter state that Xen can be used where more safety and isolation is required, and Linux-Vserver can be used in other circumstances to reduce overhead. This decision reflects the fundamental conflict between the design goals for isolation described earlier in this section.

3.2.4   APPLICATION VIRTUALIZATION

Application virtual machines run as processes within a host operating system. An application VM typically provides a platform-independent programming abstraction that hides external dependencies such as the host operating system and libraries. Such VMs are typically associated with specific programming languages; for example, Java virtual machines execute Java bytecode and Microsoft's .Net Common Language Runtime (CLR) executes bytecode generated from one of several programming languages, including C#. To deal with state external to the virtual machine or to improve performance, some application code is executed *natively* by invoking methods that are expressed not as bytecode but rather in the executable format required by the platform on which the application VM is running. The application VM typically partitions application data into virtual state manipulated by bytecode execution and native state manipulated by native functions. Native functions and state are often particularly challenging for cyber foraging systems because the binary format may differ between the local computer and the surrogate. Native functions also provide the gateway through which code running in an application VM interacts with state external to the VM; use of such functions may often need to be restricted to guarantee result-equivalence.

CloneCloud uses the Dalvik VM, which is included as part of the Android platform, for isolation. This choice provides excellent transparency for Android programs. Both mobile computers and surrogates run the Dalvik VM, so the same bytecode executes on local and surrogate platforms without any programmer effort.

CloneCloud migrates functionality to surrogates at thread granularity. When its partitioning engine decides to move a thread from the mobile computer VM to the surrogate VM, CloneCloud first suspends the target thread using the thread suspension mechanism provided by the Dalvik VM. It then transfers the thread's state. The state consists of the per-thread stack, the thread's register contents, and the heap objects that will be needed during remote execution. While identifying all such heap objects would be very difficult if the migrated thread were written in an unmanaged language such as C++, the type safety of languages such as Java makes it possible to compute the transitive closure of all references starting with the local data objects in the stack frame.

CloneCloud next accounts for differences in the platform-specific representation of each migrated object. It converts object representations to adjust for differences in byte-ordering between processor architectures and replaces non-portable native pointers to class methods with platform-independent class and method names.

Since CloneCloud may migrate some but not all threads from the mobile computer to the surrogate, application execution can sometimes proceed in parallel on the mobile computer and surrogate. However, when a mobile computer thread accesses an object that has been migrated to the surrogate, it blocks until the migrated thread and its associated object returns to the mobile computer in order to ensure that all data accesses are consistent with results that could have been generated on a single machine. The CloneCloud authors speculate that a distributed shared memory implementation might improve performance by replacing blocking with additional network communication to synchronize state across platforms.

When its partitioning engine decides to migrate a thread running on the surrogate back to the mobile computer, CloneCloud employs a similar mechanism, called reintegration, to move state updated on the surrogate to the mobile computer. CloneCloud maintains a mapping between each object on the surrogate and the original object on the mobile computer from which it was cloned. During reintegration, CloneCloud creates new objects on the mobile computer to reflect any objects created during remote execution on the surrogate, deletes any mobile computer objects that correspond to objects garbage collected during remote execution, and copies the contents of remaining objects from the surrogate to the corresponding mobile computer object. The state merge preserves result-equivalence because CloneCloud has blocked any mobile computer thread before it could modify the contents of a mobile computer object while the object was being accessed remotely. In other words, there can be no conflicting updates to objects.

MAUI uses Microsoft's CLR for isolation. CLR programs are compiled from languages such as C# into the CIL intermediate language. At run-time, CIL is dynamically compiled to run on different platform-specific architectures. Like CloneCloud's uses of the Dalvik VM, MAUI's use of CLR provides strong transparency: programmers do not have to make their code platform-aware because such details are managed by the application virtual machine.

Offloading in MAUI is done at method granularity. For each method that might be executed remotely, MAUI generates a wrapper function with two additional parameters that are used to send state to the remotely executed method and return results. MAUI takes advantage of the type safety of the .NET runtime by traversing the objects used by the program to determine the set of state that may be accessed by the remotely executed method. This includes local objects, including those with nested, complex types, as well as static classes. MAUI uses support provided by .NET for XML-based serialization to transform state into a machine-independent representation for transfer between the mobile computer and surrogate. As an optimization, MAUI reduces the amount of state that must be transferred by shipping incremental deltas of modified application state rather than the entire state when possible.

As discussed above, application virtual machines typically provide excellent transparency because developers need not modify their applications to run on different hardware platforms. For the remaining four goals outlined previously in this section, application virtual machines occupy a middle point in the design space between process-level isolation and hardware virtual machines.

Most application VMs guarantee result-equivalence for bytecode that does not execute any native functions: the virtualized computer that executed that bytecode remains the same even if the physical computer on which the VM executes changes. If an application VM does not provide complete safety guarantees, it is often possible to add additional enforcement mechanisms to enforce result equivalence. For instance, the Scavenger cyber foraging system [Kristensen, 2010] is based on mobile Python code that executes in an execution environment written using Stackless Python. The Python environment does not provide safety properties as strong as those provided by the Dalvik VM or the .NET CLR. To compensate, Scavenger declares language features such as many of the reflective feature of Python to be dangerous. Such code features are black-listed so that they can not be executed by the mobile Python code. The restricted execution is therefore guaranteed to be result-equivalent.

Application virtual machines do not provide result-equivalence for native functions. However, since the interface to native code is well delimited, it is possible to restrict the use of such functions on surrogates. MAUI's execution model does not allow remote execution of code that interacts with external entities, such as hardware devices that exist only on the mobile computer, external network interfaces, and the user interface. If a method executing remotely attempts to invoke a method that performs one of the above actions, MAUI transfers control back to the mobile computer. CloneCloud allows some native functions to execute both on the surrogate and the mobile computer. However, any method that executes a feature specific to the mobile computer must execute on that computer, and methods that share native state must execute at the same location. Scavenger white-lists the Python module imports that can be used remotely. All of the above approaches require some manual analysis by developers to determine which functionality is safe to execute remotely and which functionality is not. However, such analysis is typically not application-specific since it pertains to the interface between code executing in the VM and native functions. Therefore, the analysis needs to be done only once by the developers of the cyber foraging system; no additional analysis is required from application developers.

Application virtual machines limit flexibility since they typically execute code written in a specific programming language (or a small set of languages). From the perspective of any particular mobile computer, this limitation may not be severe. For instance, Android applications are already written in Java to execute in the Dalvik VM, so the flexibility of application writers is not additionally restricted by limiting code to run in the Dalvik VM on the surrogate. However, the diversity of application VMs used across mobile platforms makes surrogate deployment more challenging since a surrogate would need to support all such environments to handle every nearby mobile computer (e.g., Apple, Android, and Windows phones would all require different application VMs).

The overhead and start-up time for an application VM are both typically less than required for a hardware VM because the application VM contains less state; for instance, the application VM does not include a guest operating system. On the other hand, the overhead and start-up time for an application VM will be greater than the overhead and start-up time when using process-level isolation. Since each application VM runs inside a separate process, the base cost of creating the VM is at least as great as the base cost of creating a process. In addition, the application VM adds functionality such as code loading, just-in-time compilation, interpretation, and garbage collection. These features impose a performance penalty both during initialization of the VM and while the VM is executing application code.

Figure 3.2 summarizes the discussion in this section by comparing the three isolation mechanisms that have been discussed: process-level isolation, application virtual machines, and hardware virtual machines. The table attempts to provide a relative comparison among the three methods for each design goal discussed in this section. While it may be possible to quibble with the individual labels in the table, the overall results show why choosing an isolation mechanism is difficult: no one

	Process Isolation	Application VM	Hardware VM
Representative Systems	RPF Spectra Chroma	MAUI CloneCloud Scavenger	Slingshot Cloudlets
Safety	Poor	Moderate	Excellent
Flexibility	Limited by surrogate environment	Limited to specific language	Very flexible
Transparency	Poor	Excellent	Excellent
Overhead	Minimal	Moderate	Moderate
Start-up Time	Minimal	Moderate	Substantial

This table summarizes the tradeoffs among the three methods for isolating surrogate operations that have been discussed in this section. The first row gives examples of cyber foraging systems that have used each isolation mechanism, and the next five rows provide a general comparison for different design goals.

Figure 3.2: Comparison of isolation mechanisms.

option clearly dominates the other along all design axes. Improving existing isolation mechanisms to better meet the needs of cyber foraging would therefore seem to be a promising area for further research.

3.3   MANAGING STATE

Operations that execute on surrogates require two types of state. The first is the common state that is part of the surrogate infrastructure. Examples of common state are the host operating system and software libraries installed on the surrogate, a hardware virtual machine monitor, or application VM software such as the Dalvik VM or the Common Language Runtime. The common state is the lowest common denominator for remote execution: it represents the set of functionality that a particular cyber foraging system can assume about a surrogate to which it sends remote operations. Common state is persistent; it is installed when the surrogate is first deployed, and it is upgraded according to good software maintenance practices. The common state is also logically read-only; since it is shared by many mobile computers, no remote operation should perturb the state in a manner that would be visible to operations performed by another mobile computer.

3.3.1   APPLICATION-SPECIFIC STATE

The second type of state is application-specific. The previous section discussed how many cyber foraging systems ship the in-memory data that a particular remote operation will use as input to surrogates at the beginning of each operation. Such input parameters are part of the application-specific state. Another part of the application-specific state is the source code, bytecode, or binary that executes the remote operation. Some cyber foraging systems allow remote operations to access data stored in the file system; in that case, the accessed files are another part of the application-specific state. In general, we define the application-specific state to be any data required for a remote operation on a surrogate that is not part of that surrogate's common state.

Whereas the common state is persistent and read-only, the application-specific state is both transient and mutable. For some cyber foraging systems, application-specific state persists on a surrogate only for a single remote operation. Other systems have a connection abstraction in which a mobile computer connects to the surrogate, performs several remote operations, and eventually disconnects. Such systems may choose to allow application-specific state to remain on the surrogate for the duration of a single connection. Still other cyber foraging systems store application-specific state between connections, to reduce start up costs if the same mobile computer connects to the surrogate again in the future.

The decision about how long application-specific state may persist on a surrogate is often driven by the size of the state. Cyber foraging systems that isolate remote execution in hardware virtual machines have large amounts of application-specific state. The start-up cost of transferring needed state to the surrogate is substantial. However, the start-up cost can be amortized across many remote operations if the state can persist on the surrogate for many such operations. Thus, cyber foraging systems that use hardware VMs typically allow state to persist for an entire connection or even across multiple connections. On the other hand, cyber foraging systems that use process isolation or application virtual machines require less application-specific state to perform an operation. Such systems typically ship the application-specific state at the beginning of each operation or allow the state to persist for only a single connection.

Since application-specific state is logically private to each mobile computer, surrogates that store such state generally allow remote operations to modify such state. However, most cyber foraging systems require that all such application-specific state be *soft state*. In other words, the surrogate does not guarantee that it will reliably store a copy of the state; the state may be destroyed due to a surrogate fault, or the surrogate may simply decide to discard the state if it is running low on resources. For instance, cloudlets store only soft state such as cached copies of data or code that are available elsewhere. Therefore, the loss of the cloudlet incurs only a performance impact; it does not decrease the reliability of the applications that the cloudlet is hosting.

Slingshot is an exception to the "no soft state" rule-of-thumb. Recall that Slingshot assumes that some surrogates, called home servers, are first-class replicas hosted in the cloud that have resources persistently dedicated to a mobile computer. These reliable first-class replicas store hard state. However, Slingshot also assumes that some surrogates are second-class replicas deployed opportunistically at third-party sites near mobile computers. Slingshot does not assume that second-class replicas are reliable, and so, like other cyber foraging systems, requires them to store only soft state.

This discussion indicates that the prohibition against storing hard state on most surrogates implicitly derives from the transitory nature of the relationship between mobile computers and surrogates. If there exists a dedicated surrogate for the user in the cloud, then it is reasonable to assume that appropriate data replication and backup schemes are available to safely store hard state. If surrogates are discovered and used opportunistically, then mobile computers may not trust the reliability of third parties managing those surrogates. It is safer therefore to ensure that no state is permanently lost if such a surrogate fails.

3.3.2   OBTAINING AND CACHING APPLICATION-SPECIFIC STATE

We next describe several models currently used by cyber foraging systems to deliver application-specific state to surrogates. We also explore the mechanisms used to cache such state on surrogates between operations or connections.

Direct transfer to surrogate

The simplest method to push application-specific state to a surrogate is to send the state in its entirety prior to beginning an operation. CloneCloud uses this method. The application-specific state consists of a thread and the data upon which the thread may operate. Since the amount of application-specific data is relatively small, the entire state can be sent from the mobile computer to the surrogate during each thread migration. MAUI uses a similar method to transfer application data objects. However, MAUI caches state on surrogates between operations, so it can optimize subsequent transfers by shipping a delta containing only state that has changed.

Objects in application virtual memory are usually transferred to the surrogate directly from the mobile computer because the mobile computer has the most up-to-date versions of such data. For other portions of application-specific state, it may make more sense to retrieve data from other locations. For instance, code and binaries are relatively static, so the cyber foraging system might ask the surrogate to retrieve them from a data repository in the cloud. MAUI surrogates can either fetch CIL executables directly from a mobile computer, or the mobile computer can specify a signature of the executable that the surrogate will use to download the corresponding executable from a cloud service.

Goyal and Carter's cyber foraging system extends the cloud model by having mobile computers specify a URL that points to a program, typically a shell script, that the surrogate downloads and executes. The program may download and install additional software package and further customize the execution environment.

Content-addressable storage

Many cyber foraging systems cache application-specific state on surrogates to reduce the start-up time for operations that can re-use such state. For instance, Goyal and Carter's system lets mobile computers save the state of root partitions after they have been customized by running the setup script. Such caching is typical for systems that use hardware virtual machines, because the persistent state of the VM images is so large that transferring the entire state on each remote operation is prohibitively expensive.

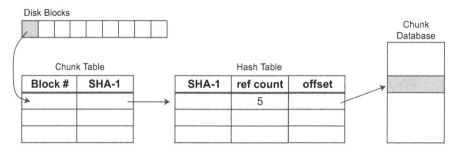

Figure 3.3: Slingshot: reading a data chunk from a content-addressable store.

Some cyber foraging systems use *content-addressable storage* to take advantage of similarities between the state used by different mobile computers. A content-addressable store breaks the application-specific state into smaller chunks, each of which is indexed by a secure hash value calculated via an algorithm such as SHA-256. Ideally, the properties of the secure hash function ensure that the likelihood of two different chunks hashing to the same value is vanishingly small and that it is computationally infeasible to generate a data chunk that hashes to the same value as another chunk. If two application-specific states are similar, then they are likely to share many data chunks. For instance, two similar VM images might share the operating system and library code, as well as the code of many utilities and applications.

A content-addressable store identifies cases where two or more data chunks are the same by comparing the hash values of all chunks. If two or more chunks hash to the same value, the store saves only a single physical copy, thereby conserving storage capacity. Content-addressable stores can also help reduce the amount of data that needs to be transferred when sending new application-specific state to the surrogate. Data chunks that are already present on the surrogate can be identified by their hash values; only chunks not present are sent to the surrogate.

Slingshot is an example of a system that uses a content-addressable surrogate cache. Slingshot maintains a virtual machine image for each application. This image resides permanently on the dedicated first-class replica on the home server. Additional second-class replicas are instantiated on demand at surrogates located near the mobile computer in order to improve the performance of remote operations.

Slingshot adopts prior content-addressable storage techniques [Sapuntzakis et al., 2002; Tolia et al., 2004] by dividing the disk image of each virtual machine into 4 KB chunks and indexing each chunk by its SHA-1 hash value. As shown in Figure 3.3, each virtual machine has a *chunk table* that maps the chunks in its virtual disk image to the SHA-1 hash of the data stored at each location.

The home server maintains a hash table of the SHA-1 values of all chunks that it currently stores. When it receives a request to store a new chunk whose SHA-1 value matches that of a chunk it already has stored, it increments a reference count on the existing chunk. This method of eliminating duplicate storage has been shown to substantially reduce disk usage [Cox et al., 2002] due to similarities between the disk state of different computers. Slingshot expects such similarities to be common because a single user is likely to create many application-specific VMs from the same generic OS image.

As shown in Figure 3.3, when a first-class replica running on the home server reads a chunk from its virtual disk, Slingshot intercepts the I/O request and looks up the block number in the service's chunk table to determine the SHA-1 value of the chunk stored at that location. It then looks up the SHA-1 value in the hash table to find the location of the chunk in the chunk database.

Requests that modify blocks follow a similar path. Slingshot locates the chunk in the service's chunk table. It then indexes on the old SHA-1 value and decrements the reference count associated with the chunk in its hash table. If the reference count drops to zero, it deletes the chunk. Slingshot next looks up the new SHA-1 value of the modified block in its hash table. If the modified chunk is a duplicate of an existing chunk, it increments the reference count of the existing chunk. Otherwise, it stores the chunk and adds its SHA-1 value to its hash table.

Slingshot uses a similar content-addressable store for the disk images of virtual machines instantiated opportunistically on surrogates located near the mobile computer. Since such surrogates are not dedicated to specific users, the content-addressable store is only a cache.

As with the home server, second-class replicas also index each chunk of a virtual machine's disk image by its SHA-1 hash value and eliminate the storage of duplicate chunks. This lets users benefit from similarities among the disk images of their replicas. For instance, two people using Windows-based services may have similar disk images. Chunks cached by one user can be used by the other.

When a virtual-machine running as a second-class replica reads a chunk, the surrogate first checks to see if the chunk is present in its local cache. If so, it returns the associated data immediately. If the chunk is not present, the surrogate fetches the chunk from the home server of the user whose virtual machine requested the chunk. It caches the returned chunk and also returns the contents of the chunk to the virtual machine that requested the data.

As an optimization, a surrogate hosting a second-class replica may fetch chunks from the nearby mobile computer [Su and Flinn, 2005b]. When the mobile computer is well connected to the home server, it downloads chunks into a local content-addressable store. When the mobile computer establishes an opportunistic connection to a new surrogate, it uploads its hash table to the surrogate. Later, when the surrogate needs a chunk that is not in its cache, it fetches the chunk from the mobile computer if there is a matching hash value in the uploaded hash table; otherwise, it fetches the chunk from the home server. Servicing cache misses from the mobile computer can substantially improve performance because the local network connection between the mobile computer and surrogate is typically much better (higher bandwidth and lower latency) than the connection between the surrogate and the home server.

Note that in content-addressable storage systems, chunks are self-validating. If data stored in a chunk is modified between the time when the chunk was cached on the mobile computer and the time when it is requested by the surrogate, then the hash value of the chunk (which depends on the chunk contents) will change. The new hash value will not match the old value for the cached chunk and thus will not be found in the mobile computer's hash table. Therefore, the correct contents of that chunk will be fetched from the home server.

Slingshot's surrogate cache uses an LRU eviction algorithm. Since chunks remain cached even after a service is terminated, it is likely that the chunks belonging to a mobile computer that is frequently near the same surrogate will remain cached between visits.

VM Synthesis

An alternative method for generating application-specific virtual machine images on surrogates is *VM synthesis*. This technique, used by cloudlets, assumes that surrogates maintain a *baseVM*, which contains a generic, minimally configured guest operating system. Mobile computers instantiate new application-specific VMs by shipping the surrogate a compressed delta of the differences between the desired application-specific VM image and the baseVM. The surrogate applies the overlay to instantiate the custom VM required by the mobile computer.

For the first generation of cloudlets, the overlay was created by a tool called `kimberlize` that instantiated a well-known baseVM, executed an install-script to customize the software environment inside the VM, and ran a resume-script to start services inside the VM and bring them to a state in which they are ready to receive requests from a mobile computer. The `kimberlize` tool next calculated the difference between the newly created VM state and the baseVM state and applies compression to create the overlay. Thus, the process of creating a cloudlet overlay was somewhat akin to the customization of a VM in Goyal and Carter's cyber foraging system; however, the `kimberlize` tool was run once, to statically create the overlay, whereas Goyal and Carter's system runs the customization script every time a new VM is instantiated. The current cloudlet work has moved away from the `kimberlize` model of overlay creation and is instead making cloudlet support an extension of OpenStack.

In the original instantiation of the cloudlet model, the overlay was sent from the mobile computer to the surrogate. This realizes the same benefits achieved by Slingshot when VM state chunks are sent from the mobile computer to the surrogate: the local, one-hop network connection will typically have higher bandwidth and lower-latency, reducing VM start-up time. VM instantiation using overlays has been shown to take less than 2 min [Satyanarayanan et al., 2009]. The effect on battery lifetime of sending overlays from the mobile computer instead of the cloud is unclear, however. Reducing start-up time saves battery energy, but the mobile computer may expend more energy to read custom VM state from storage and send it to the surrogate. For this and other reasons, more recent instantiations of the cloudlet model can obtain the overlay from either a mobile computer or the cloud.

Leveraging distributed storage

A final model to deliver application-specific state to a surrogate is for the mobile computer and surrogates to place such data in a distributed storage system shared by mobile computers and surrogates. For example, Spectra stores executables, libraries, and the data accessed during remote operations in the Coda distributed file system [Kistler and Satyanarayanan, 1992].

To guarantee result-equivalence, file system operations performed on surrogates must produce the same result that would have been produced if the same operations had been performed on the mobile computer. For this reason, prior remote execution systems such as Butler [Nichols, 1987] used a distributed file system that presented a consistent file system image across all machines on which functionality may be executed.

Unfortunately, file consistency comes at significant cost in mobile computing environments, in which network connections to file servers often exhibit high latency, low bandwidth, or intermittent connectivity. When an application modifies files, it may block for long periods of time waiting for data to be reintegrated to file servers.

Those performance consideration led Spectra to use Coda. Coda provides strong consistency when network conditions are good. Under low bandwidth conditions, Coda buffers file modifications on the mobile computer to improve performance. Buffered modifications are reintegrated to file servers in the background. Until a modification is reintegrated, it is not visible on other machines.

Spectra interacts with Coda to provide the requisite consistency for remote execution. Before it executes functionality remotely, Spectra predicts what files will be accessed by the application. If any such files have buffered modifications on the mobile computer, Spectra triggers their reintegration to file servers before executing the remote operation.

The advantage of using a storage system to distribute application-specific state is simplicity—the complexity of managing cache consistency and dealing with poor wireless networks is managed by a lower system layer. The disadvantage of this solution is that the strategies chosen by the storage system to propagate data may not be optimal for the needs of the cyber foraging system. For instance, when the newest copy of data exists only on the mobile computer, the data must first be sent to the file server (over a potentially high-latency, low-bandwidth network link) and then back to the surrogate. A one-hop local network transfer will typically yield better performance.

3.4   MECHANICS

The prior sections in this chapter have discussed issues pertaining to how surrogates are deployed and how application-specific state is pushed to surrogates and potentially cached for later use. In this section, we discuss the mechanics of how a mobile computer discovers nearby surrogates, connects to one of the available surrogates, and performs remote operations.

Cyber foraging clients "live off the land" by dynamically discovering surrogate computers in their immediate network neighborhood. Even Slingshot, which leverages well-known home surrogates in the cloud, supplements the home surrogate by using opportunistically discovered surrogates located near the mobile computer. Thus, a surrogate discovery protocol is an important part of any cyber foraging system.

Goyal and Carter provide a custom discovery service for their cyber foraging system. Surrogates register with a well-known surrogate discovery server by providing an XML-like description of their capabilities. This information can include the operating system and other software supported by the surrogate. Mobile computers can discover suitable surrogates with XML-like queries. Mobile computers are expected to provide a conservative estimate of their resource requirements, which can be used, for example, to identify surrogates that have enough spare capacity to service requests promptly from the mobile computer. The surrogate discovery server responds to the mobile computer with the IP address of a surrogate that meets the mobile computer's needs. It also specifies the port number on which the surrogate's manager process is listening — the mobile computer contacts the manager to install and start using a virtual machine.

Other cyber foraging systems have chosen to use common-off-the-shelf service discovery protocols. Cloudlets use the Avahi zero configuration networking implementation to discover surrogates available on the local wireless network. Spectra uses Universal Plug and Play (uPnP), which is a set of networking protocols that permits networked devices to discover one another. Scavenger uses a service discovery framework called Presence.

If more than one surrogate is available for a mobile computer to use, the cyber foraging system must choose the best surrogate(s). In Goyal and Carter's system, the surrogate discovery server makes this decision. Since it has both the mobile computer's resource request and the available surrogates' capabilities, the server is best suited to make a good match. In other systems such as cloudlets and Spectra, the mobile computer must choose between surrogates if more than one is available. However, since the service discovery protocols used by these systems are limited to a local network, it is unlikely that more than one surrogate will be discovered.

While most cyber foraging systems utilize a single surrogate at a time, Chroma and Slingshot can opportunistically make use of multiple surrogates. As discussed in the previous chapter, Chroma can execute operations on more than one surrogate in an over-provisioned environment in order to reduce the time taken to complete remote operations or to provide the best fidelity within a specified time limit. However, Chroma does not specify a specific policy for deciding when to employ redundancy.

Slingshot uses multiple surrogates to hide the start-up cost of installing application-specific state on new surrogates. Slingshot assumes that the surrogate on the home server is continuously available. When the mobile computer connects to a new network, it searches for a surrogate co-located with the hotspot. If such a surrogate exists, the Slingshot client instantiates application-specific state on that surrogate, as described in the previous section. This may be a lengthy process, but the Slingshot server can use the home server to execute remote operations while the nearby surrogate is initializing.

If the remote operations are logically read-only (i.e., they do not modify the application-specific state), then the new surrogate is initialized as soon as it has obtained the transient memory

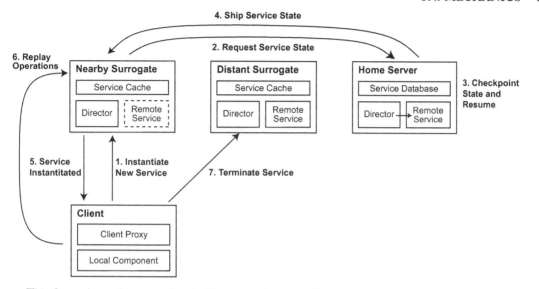

This figure shows the steps taken by Slingshot when a mobile computer moves from one hotspot to another. Slingshot first instantiates an additional replica at the new hotspot, then it terminates the replica at the old hotspot.

Figure 3.4: Instantiating a new Slingshot replica.

state of the virtual machine and the chunk table that indexes the VM's persistent state. However, if remote operations executed during initialization modify the application-specific state, then the initialized state will be out-of-date. Essentially, the state transferred to the new surrogate is a stale checkpoint of the state of the replicated VM. The Slingshot client therefore logs any state-modifying operations that occur while a surrogate is being initialized. After initialization completes, it replays all logged operations to bring the new surrogate up-to-date. Once there are no more operations in the log, the Slingshot client broadcasts all new operations to both the home server and the new surrogate.

The benefit of replication is that the user sees little foreground performance impact due to the use of a new surrogate. After checkpointing, the first-class replica on the home server services requests while the new second-class replica is instantiated and brought up-to-date. In contrast, migrating the VM from the home server to the new surrogate would make the remote service unavailable while state is being shipped.

Figure 3.4 shows how Slingshot responds to a mobile computer moving to a new hotspot. While Slingshot is initializing the nearby surrogate at the new hotspot, it continues to broadcast operations to both the distant surrogate at the previous hotspot and the home server. Note that Slingshot does not need to determine which replica will perform best since it simply uses the fastest response — this strategy is akin to Chroma's broadcast of remote operations to multiple surrogates.

When the nearby surrogate is ready to accept new operations, the Slingshot client disconnects from the surrogate at the distant hotspot.

Returning to the general mechanics of using surrogates, after a mobile computer has discovered an initialized a surrogate, it must establish a network connection to that surrogate. Most surrogates have a management service listening on a well-known port. Mobile computers first connect to this manager to set up an environment that they can use to perform remote operations. The manager may perform admission control. For instance, in Goyal and Carter's system, the surrogate manager either establishes a contract with the mobile computer to provide a fixed amount of service for a specified duration, or it rejects the connection. The manager and the mobile computer may also mutually authenticate — we defer discussion of this step until the next chapter, in which security and privacy issues are discussed.

Section 3.3 described the different methods in which the surrogate manager and the mobile computer may cooperate to set up application-specific state on the surrogate. This state may be as simple as a process running a specific server, or it may be a custom hardware virtual machine. Typically, during this setup, the mobile computer learns of one or more ports on which server processes within that custom state are listening for incoming connections. It sends requests to perform remote operations to those ports.

Mobile computers may explicitly disconnect from a surrogate by sending a message to the surrogate manager. Due to the vagaries of wireless networking in which mobile computers may unexpectedly become disconnected, surrogates can never assume that clients will always explicitly disconnect. Therefore, the surrogate manager must also use a timeout or similar method to implicitly detect dropped connections. When a mobile computer disconnects, the surrogate will typically stop any processes running on its behalf. The surrogate may deallocate any application-specific state associated with that mobile computer immediately, or it may decide to cache that state to improve start-up time if the same computer reconnects in the future.

3.5   SUMMARY

This chapter discussed several important design options in the deployment and operation of surrogates. Section 3.1 described the tension between locating surrogates in centralized data centers and positioning them as close to mobile computers as possible. Most cyber foraging systems to date have advocated for dispersing surrogates to reduce communication time between the surrogate and the mobile computer. Slingshot tries to achieve the benefits of both locations by using multiple replicas.

Section 3.2 explored options for isolating state on surrogates. Strong isolation makes it easier for a cyber foraging system to guarantee result-equivalence, the property that the observable results of an operation that executes remotely on a surrogate should be indistinguishable from results that could have been produced by the same operation if it had executed on the mobile computer. The method for providing isolation also impacts the flexibility, transparency, performance overhead, and start-up time for executing remote functionality on surrogates. The earliest cyber foraging systems generally used process-level isolation provided by the operating system, which provided only limited transparency and result-equivalence. More recent cyber foraging systems have tended to leverage advances in virtualization by running remote operations in either hardware or application virtual machines.

Section 3.3 discussed methods for transferring application-specific state to a surrogate. Application memory state is typically transferred at the start of each remote operation. Persistent file system state is larger and transferring the state in its entirety before an operation may introduce unacceptable delays. Some cyber foraging systems use content-addressable storage for such state or dynamically synthesize the state from a generic template. Other cyber foraging systems defer the management of such state to a lower-level distributed storage system.

Section 3.4 discussed the typical mechanics whereby a mobile computer discovers one or more surrogates, selects a surrogate on which to perform remote operations, establishes a connection to that surrogate, executes operations, and terminates the connection.

The work in cyber foraging systems to date, and hence the discussion in this chapter has focused mostly on technical design of the surrogate infrastructure. Because cyber foraging is still in its infancy, there has been relatively little work on the management of surrogates and economic issues related to their deployment. This is understandable since a realistic investigation of such issues will require a concrete usage model that may only be observed once a substantial deployment of surrogates exist. Such future challenges may include load management, software maintenance, cost models, and pricing.

CHAPTER 4

Security and Privacy

This chapter explores the security and privacy challenges that arise from supporting cyber foraging. Several challenges such as secure communication between the mobile computer and surrogates are standard challenges in other domains. Primarily, the challenges unique to cyber foraging come from executing arbitrary computation on shared surrogates owned and managed by third parties with whom the owner of a mobile computer may not have a pre-existing trust relationship. This chapter will therefore focus on these specific issues.

The next section begins by listing the desirable security and privacy properties of a cyber foraging system. Surrogate providers would like to disallow remote operations executed on a surrogate from performing malicious actions. Surrogate providers would also like to ensure that multiple tenants on a single surrogate cannot interfere with one another. On the other hand, mobile computer owners desire that surrogates perform remote operations correctly and that surrogates keep data associated with the operations confidential.

Section 4.2 argues that several of the above properties are already concerns in existing hosting environments such as multi-tenant cloud data centers. Therefore, current security mechanisms, such as virtual machine isolation, can provide some of the needed solutions. However, cyber foraging introduces unique security and privacy challenges because the relationship between a mobile computer owner and a surrogate provider may be transient and because the public deployment of surrogates increases the opportunity for physical tampering. Thus, cyber foraging users may reasonably consider operations hosted by a surrogate to be less secure than operations hosted by a multi-tenant cloud data center.

The next two sections therefore describe additional solutions that enhance the trustworthiness and confidentiality of executing remote operations on surrogates. Section 4.3 discusses how a trusted boot process can give mobile computers a guarantee that the software running on a surrogate is trustworthy. Section 4.4 details methods for verifying the correctness of surrogate computation by executing redundant computations on trusted computers.

4.1 DESIRED SECURITY AND PRIVACY PROPERTIES

There are two primary classes of participants in cyber foraging systems: mobile computer owners that wish to perform certain operations on remote infrastructure and surrogate providers that deploy infrastructure to host such computation in exchange for monetary or other benefits. Each class of participant desires different properties from the cyber foraging system. Malicious third parties may wish to disrupt the cyber foraging system (denial of service), discover sensitive information used or produced by a remote operation, or cause a remote operation to produce incorrect results.

4.1.1   PROPERTIES FOR SURROGATE OWNERS

We first consider the needs of surrogate owners. Surrogate providers wish to ensure that remote operations hosted on surrogates perform no malicious or harmful actions. For instance, a remote operation should not be able to compromise the software infrastructure of the surrogate so as to perturb or eavesdrop on the operations of other mobile computers. In practice, it is infeasible to enumerate all possible malicious or harmful actions. Therefore, the surrogate owner must be content with a slightly weaker property in which a service-level agreement or other specification partitions potential actions into those that are allowed and those that are disallowed and the cyber foraging system prevents a remote operation from performing any disallowed action. Ideally, the list of disallowed actions should severely restrict the ability of a hosted operation to cause substantial harm.

Surrogate owners also wish to prevent remote operations from one or more mobile computers from degrading the quality of service provided to other computers using a surrogate. The remote operations performed by some mobile computers may request a substantial amount of surrogate resources, either because the operation is naturally resource-intensive or because the code describing the operation is buggy. Alternatively, a malicious party may attempt a denial of service attack by trying to consume so many resources that other mobile computers can make little progress.

4.1.2   PROPERTIES FOR MOBILE COMPUTER OWNERS

Mobile computer owners primarily wish to ensure the trustworthiness and confidentiality of remote operations. A remote operation is defined to be trustworthy if it faithfully performs its intended purpose. In practice, trustworthiness is a somewhat difficult property to define precisely. One reasonable definition relies on the *result-equivalence property* defined in the previous chapter: the observable results (including side effects) of an operation that executes remotely on a surrogate should be indistinguishable from results that could have been produced by the same operation if it had executed on the mobile computer. This definition assumes that the mobile computer is itself trustworthy (and therefore performs operations correctly).

Consider a surrogate that provides result-equivalency. If a particular remote operation is deterministic, every invocation of that operation with the same inputs should produce only the same result. The result produced by the surrogate will be the same as that produced by a trustworthy mobile computer; the result is therefore correct. If a remote operation is non-deterministic, there exists a set of possible results that may be produced by the mobile computer given the same input. A surrogate that is result-equivalent will always produce one of these results. Although a malicious surrogate may produce a low-probability result, that result can be considered correct because it could potentially have been produced by a trustworthy mobile computer. A stronger definition of trustworthiness could potentially include some notion of probability of results for non-deterministic operations; however, constructing such a distribution for general computation seems very challenging.

Mobile computer owners may also desire that the information associated with a remote operation remain confidential. Confidentiality is a challenging property to provide since the surrogate hardware and software (virtual machine monitors and operating systems) must read information in order to perform computations on the data. Such entities may directly leak sensitive information, or they may indirectly communicate the data via hidden channels. While confidentiality guarantees remain elusive in cyber foraging system, trusted boot mechanisms, discussed in Section 4.3, provide some protection against malicious software.

4.2   LEVERAGING STANDARD SOLUTIONS

From the previous discussion, it can be observed that the security needs of surrogate providers are similar to those of the operators of multi-tenant cloud data centers. Cloud data center operators commonly virtualize physical resources such as servers and storage to reduce costs [Calder et al., 2011]; they execute multiple virtualized computations on the same physical server to fully utilize the computing power of each server in the data center. This model of sharing the computational resources is similar to that of a surrogate that hosts operations requested by multiple computers.

Virtual machine sandboxing is the standard security mechanism used by cloud data center operators to prevent hosted operations from performing malicious actions. Cloud data centers also typically implement performance isolation by capping the maximum resources that each virtual machine can acquire within a given time period. Since such standard methods should work equally well for surrogate operators, many cyber foraging systems incorporate these techniques — in fact, the details have already been discussed in the sandboxing discussion in Section 3.2.

However, surrogate providers face additional complications not seen by cloud data center operators. First, there will often be no permanent relationship between surrogate and mobile computer owners; a mobile computer may use a given surrogate only once and such opportunistic use may have little lead time. The lack of a long-term relationship makes it harder to establish strong identities and use social or legal mechanisms to enforce good behavior. This may be alleviated if a given provider operates a large network of surrogates in many locations. Such a provider could require a mobile computer owner to register once with identifying information before using any surrogate. The mobile computer could then provide identifying information obtained during registration before using any surrogate.

Physical tampering is a much greater threat for surrogates than for servers in cloud data centers. Whereas the set of people that can access servers in a data center can be easily restricted, it may be much harder to constrain the set of people that have physical access to surrogates, which are deployed at or near public locations.

Standard security solutions can also be leveraged by cyber foraging systems to establish secure communication between mobile computers and surrogates. For example, Goyal and Carter's system [Goyal and Carter, 2004] uses such mechanisms by assuming that a pre-existing trust relationship exists between a mobile computer and a surrogate. Each surrogate maintains a list of the public keys of computers authorized to perform remote operations on the surrogate. Mobile computers use SSL/TLS to communicate with the surrogate manager. The surrogate manager creates virtual servers to handle the mobile computer's remote operations. It configures each virtual server for a particular computer by including the mobile computer's public key in the root SSH configuration authorized keys file. This allows the mobile computer to use SSH to communicate with its virtual servers. Goyal and Carter's system also includes a layer of indirection in which surrogates maintain user public keys, and users can authorize multiple computers to use the surrogate on their behalf.

The surrogates in Goyal and Carter's provide only self-signed certificates to mobile computers; thus, they do not authenticate themselves to mobile computers before they are used. This weakness can be addressed by leveraging public key infrastructures. For instance, a small provider might obtain a surrogate certificate from a public key authority. Alternatively, a large-scale provider that maintains a network of surrogates could sign surrogate certificates with its private key.

While standard security and privacy solutions can address the needs of surrogate providers and ensure secure communication between mobile computers and surrogates, meeting the needs of mobile computer owners is more challenging. Ensuring that computation performed on another machine is both trustworthy and confidential is not currently a solved problem. However, the mechanisms discussed in the next two sections, trusted boot and verification via redundant execution, provide some partial guarantees.

4.3   TRUSTED BOOT

A trusted boot process uses a low-level, hardware root of trust, such as a Trusted Platform Module (TPM), to enable third parties to verify that specific software has been loaded on a computer system. A trusted boot process can help prevent attacks such as installation of spyware and keystroke loggers, as well as a malicious BIOS, virtual machine monitor, operating system, or application. The key idea is that a hardware processor can generate a cryptographic attestation of each software layer as it is loaded onto the computer system, and then provide this cryptographic proof to a third party on request. The third party can check that all loaded software is trusted. If untrusted software has been loaded, the third party can refuse to interact further with the computer in question.

Trusted boot is a promising method for providing the trustworthiness and confidentiality guarantees desired by cyber foraging systems. While there has not yet been substantial work that applies trusted boot specifically to cyber foraging, the technique has been used to solve several related problems. One such similar problem is the initialization of public computing devices or kiosks [Garriss et al., 2008]. The authors of this work envision that a mobile computer can be used, along with trusted boot, to establish trust in a rental computer in an Internet Cafe or other public location. Such rental computers are very similar to the surrogates used in cyber foraging, although kiosks are not typically shared among multiple users concurrently.

4.3.1   TRUSTED BOOT FOR PUBLIC KIOSKS

A TPM is a hardware component that provides support for cryptographic primitives. While a TPM has a variety of associated cryptographic keys, the system of Garriss et al. only makes use of the Attestation Identity Key (AIK). This is a public-private key pair; the private key is stored in protected storage internal to the TPM and is not disclosed to external entities. The public AIK is included in an AIK certificate; this key can be used by external identities to verify the contents of messages signed by the TPM. Thus, it is essential that a mobile computer wishing to use a kiosk with a TPM have the correct AIK certificate for that TPM; otherwise, a malicious party could purport to be the kiosk in question by fooling the mobile computer into using a public AIK that corresponds to a private key known to the malicious party.

In a cyber foraging system, a mobile computer could perhaps retrieve the AIK certificate from the surrogate provider through an out-of-band mechanism. However, this might hinder usability by preventing opportunistic use of surrogates. The kiosks of Garriss et al. support an alternative mechanism, first proposed by the Bump in the Ether project [McCune et al., 2006], in which a mobile phone camera is used to verify the public AIK. The kiosk owner creates a bar code that contains a secure hash of the public AIK and places the bar code on the kiosk. The user reads the bar code with her mobile phone camera or bar code scanner. The mobile phone then queries the kiosk for the full public key, and rejects the supplied key if it does not hash to the value on the bar code. This prevents a malicious party from impersonating the kiosk by intercepting network traffic and supplying a bogus public AIK. However, the malicious party could potentially also replace the bar code on the kiosk; to prevent this, the bar code may be placed within a tamper-resistant casing.

In the system of Garriss et al., a user with a mobile computer wishes to ascertain that the software on a kiosk is trustworthy and that the software will keep the user's operations confidential. The user first obtains the public AIK for the kiosk as in the previous paragraph. The kiosk also supplies a list of possible software configurations that it supports. The user selects one of these options using her mobile phone; this should be a software stack (e.g., virtual machine monitor and operating system) that the user believes is trustworthy. A list of trustworthy software mechanisms could, for example, be maintained on the mobile phone, so that the phone could show which configurations available on the kiosk can be trusted. After the user selects a configuration, the kiosk then reboots and loads the selected software.

The TPM provides the assurance that the kiosk has in fact booted the selected software configuration. The TPM stores cryptographic hashes of software components in Platform Configuration Registers (PCRs), which are initialized during boot and may not otherwise be reset. As software is loaded during the boot process, the values in the PCRs are extended with the cryptographic hashes of the software images. The Integrity Measurement Architecture (IMA) extends the trusted boot process by also storing hashes of applications and configuration files.

After the boot completes, the mobile phone sends an IMA attestation request. The kiosk replies with a measurement list that contains a manifest of the software loaded since boot, as well as a quote that contains the current PCR values signed by the TPM with the private AIK. The mobile phone can verify that the software manifest matches the selected configuration. It can also independently calculate the correct PCR values for the software stack by using the known hashes of each software image. Once the software stack is verified, the user can start interacting with the kiosk.

When the user terminates the session, the kiosk should delete any session state. To make clean up easier, the kiosks of Garriss et al. store persistent data only in an encrypted file system. The encryption key for the file system, along with any memory used by a user's virtual machine are securely erased when the session ends.

4.3.2   APPLICATION TO CYBER FORAGING AND VULNERABILITIES

Since the use case for surrogates is very similar to that of kiosks, most of the secure boot process described by Garriss et al. can be directly applied to cyber foraging systems. Both environments have the desirable property that the set of software that needs to be verified is limited. In both kiosks and surrogates, the mobile computer may need only to verify a virtual machine manager, surrogate (or kiosk) management software, and a host operating system (if one exists). Garriss et al. describe an optimization whereby a dynamic root of trust can be established to avoid trusting the BIOS [Garriss et al., 2008]. Guest operating systems and applications can be loaded by the trusted VMM as directed by the mobile computer, and so do not need to be verified prior to use. Since the trusted computing base is small and not user-specific, it is reasonable to imagine that there can exist a small set of configurations supported by surrogates/kiosks and trusted by mobile users.

One substantial vulnerability of trusted boot is that software may be compromised after it is loaded, measured by the TPM, and verified by the mobile computer. If a malicious party can use a previously unknown exploit, or the database of trusted software on the mobile phone contains stale data, then an attacker can compromise a surrogate. The kiosks of Garriss et al. limit the window of vulnerability by rebooting the kiosk prior to each new usage. Such a reboot may be infeasible in a surrogate, however, because the surrogate is shared among many users who may continuously start and stop sessions. Since rebooting a surrogate currently being used by another user is not an option, the time that a virtual machine monitor or other system software can be exploited will be greater for a surrogate than for a kiosk.

Secure boot is vulnerable to hardware attacks. An attacker could potentially insert a malicious processor [King et al., 2008] or sniff architectural buses for confidential information. Because kiosks and surrogates are deployed in public areas, it may be feasible for a third party to install compromised hardware without the knowledge of the surrogate owner. The cloudlet vision [Satyanarayanan et al., 2009] suggests that surrogates can be placed in tamper-proof or tamper-evident containers with third-party monitoring of hardware integrity to help deter such attacks. Kiosks also are vulnerable to compromise of I/O hardware. For instance, a compromised keyboard could record keystrokes to sniff passwords and credit card numbers. However, surrogates are less vulnerable to such attacks because such I/O is performed using the trustworthy mobile computer.

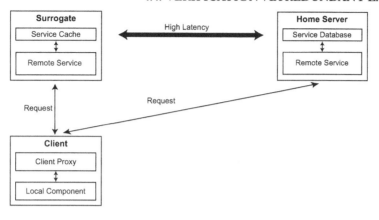

Figure 4.1: Verification in Slingshot.

4.4   VERIFICATION VIA REDUNDANT EXECUTION

Several cyber foraging have proposed or implemented verification of surrogate computation by redundantly executing some or all operations on a trusted computer. In contrast to trusted boot, redundant execution can be used only to verify that a surrogate has returned correct results from a remote operation; it does not provide any guarantee that the surrogate has kept confidential the data it observed while performing a remote operation.

The verification approaches proposed for cyber foraging systems to date differ in the type of trusted computer that is used to performs the verification. Slingshot verifies the results from all operations performed on a surrogate by also performing the same operations on the home server, which is a trusted computer in the cloud. Other systems have proposed using deterministic logging techniques to execute computation on both a surrogate and the (trusted) mobile computer. While the above systems verify all surrogate operations by executing a redundant replica on a trusted computer, it may be possible to achieve probabilistic guarantees by verifying only a random subset of operations. Systems such as Chroma that execute redundant operations on multiple surrogates may also be able to detect misbehavior if the surrogates fail independently (i.e., they are maintained by different providers and all surrogates cannot be compromised by a single attack).

We next describe some of the verification techniques used in current cyber foraging systems. We use these sample systems to explore the challenges, such as non-determinism and delayed responses, that arise from using redundant replicas to verify the correctness of surrogate execution.

4.4.1   VERIFICATION IN SLINGSHOT

In Slingshot [Su and Flinn, 2005a], a client proxy running on the mobile computer multiplexes all requests for remote operations. Figure 4.1 shows this process. The client proxy sends a copy of each request to a first-class replica on the trusted home server in the cloud. Additionally, copies of each request may be sent to one or more second-class replicas executing on nearby surrogates. Thus, each operation performed at a surrogate is also performed by the home server.

The client proxy maintains a sequential event log of all remote operations. It spawns a thread for each replica; the thread sends logged events to the replica in the order they were received. Events may optionally have application-specific preconditions that must be satisfied before they can be sent to a replica. For instance, Slingshot supports a remote desktop (VNC) application that specifies a precondition that ensures that the remote desktop is ready to accept each key stroke and mouse click event before that event is sent.

The client proxy records the replies received from each replica in the event log. The replies from all second-class replicas are compared to the reply received from the first-class replica. If the replies disagree significantly, the second-class replica with the mismatched reply is assumed to be incorrect. Consequently, Slingshot terminates execution on that replica. All other replies are verified as correct because they agree with the home server and the home server is trusted to produce correct results.

Slingshot faces two substantial challenges in performing verification. First, the home server will typically be slower to respond to remote operations than nearby surrogates. Because the home server is located in the cloud, network messages to and from the home server incur a higher round-trip delay. In addition, large messages may take longer to transmit to the home server because the last-hop uplink bandwidth to the Internet is often one or more orders of magnitude less than the bandwidth available over a single-hop, short-range wireless network such as WiFi.

The Slingshot client proxy could potentially verify each result from a surrogate prior to returning the result to the mobile application. In this case, however, Slingshot would derive no benefit from using the nearby surrogate and interactive performance would suffer due to the delay incurred by communication with the home server in the cloud. For this reason, Slingshot opts for delayed verification: it returns the first result received (usually from a second-class replica) to the mobile application but later checks that result against the (delayed) result from the first-class replica. If the client proxy returns a result that is later determined to be faulty, Slingshot notifies the application via an upcall so that corrective action can be taken. This strategy is similar to those employed in the SUNDR file system [Li et al., 2004] and in the operator undo project [Brown and Patterson, 2002]. Although not currently supported by Slingshot, it might be feasible to delay sensitive operations such as sending messages to Internet servers until all prior application operations have been verified. As long as sensitive operations are relatively infrequent, such a strategy could have minimal impact on interactive performance.

A second challenge faced in verification is non-deterministic application behavior. A non-deterministic operation may have many correct results. The particular result returned depends on non-deterministic inputs such as thread scheduling, the timing of received inputs, and random number generation. In contrast, if an operation is *deterministic*, it has no non-deterministic inputs and there exists only one correct result. For such operations, a simple value comparison as used in Slingshot is sufficient to detect faulty behavior. However, if an operation is non-deterministic, additional strategies are needed.

There are two general techniques for dealing with non-determinism. First, one can eliminate the non-deterministic behavior by ensuring that all non-deterministic inputs are the same across all replicas. Second, one can allow non-deterministic behavior but check for equivalent rather than identical results. Slingshot uses both techniques.

For remote operations with non-deterministic behavior, Slingshot uses the application pre-condition option to ensure that each operation in the log is handled sequentially. Although the proxy may have multiple operations in flight to a given replica, use of this option ensures that each replica processes the operations in the order given by the event log and that each replica does not start processing the next operation until the previous operation completes. This technique can eliminate some important forms of non-determinism such as input timing and thread scheduling. This technique can also mitigate non-determinism that arises from stateful operations because each replica sees the exact same sequence of operations.

Sequential processing of operations reduces but does not eliminate all non-deterministic behavior. For instance, the VNC server employed by Slingshot sends display updates in a non-deterministic fashion. When pixels on the screen change, it reports the new values to the mobile computer in a series of updates. Two identical replicas may communicate the same change with a different sequence of updates. The resulting screen image at the end of the updates is identical but the intermediary states may not be equivalent. A second challenge is that some applications are inherently non-deterministic. One example is that the same event (e.g., a button press) may have different results when it is delivered before and after a new window finishes displaying.

Slingshot employs an application-specific equivalency check to deal with such non-determinism. Such checks have previously been used to handle non-deterministic application behavior in replicated state machine protocols [Rodrigues et al., 2001]. Slingshot considers two replies from different replicas to differ only if the application-supplied equivalency check fails. For example, Slingshot's VNC application associates a precondition with each input event. When the user executes the event, it logs the state of the window on the mobile computer to which that event was delivered. When replaying the event on a replica, the window must be in an identical state before the event is delivered. Since each event is associated with a screen coordinate, Slingshot checks state equality by comparing the surrounding pixel values of the different execution. In the above example, this strategy causes Slingshot to wait until a window is displayed before it delivers the text entry events.

To deal with the non-determinism of updates, the Slingshot VNC application designates the best-performing replica as the foreground replica and the remaining replicas as background replicas. Only events from the foreground replica are delivered to the mobile computer. If performance changes, the client proxy quiesces the replicas before choosing a new foreground replica.

4.4.2   ENFORCING DETERMINISM

From the above discussion of the Slingshot VNC application, it is apparent that it is challenging to design a good equivalency check for complex, non-deterministic applications. Therefore, some recent cyber foraging research have proposed eliminating the need for equivalency checks during verification by comprehensively eliminating all sources of non-determinism. This ensures that correct replicas always return identical results.

One proposal [Flinn and Mao, 2011] uses *deterministic replay* to execute identical operations on both the mobile computer and a surrogate. The goal of deterministic replay is to capture an execution of a computer system so that the execution can subsequently be replayed on the same or a different computer. The success of deterministic replay is predicated on the observation that the bulk of a computer system's execution is deterministic. Thus, one need only log the small set of non-deterministic inputs and operations in order to capture an execution — these events happen at relatively low frequencies. When replaying a recorded execution, the non-deterministic operations are not re-executed. Instead, the results of those operations from the recorded execution are reproduced from the log and supplied to the replayed execution. This guarantees that the replayed execution will produce results identical to those produced by the original, recorded execution.

The main reason that deterministic replay is used in this work is to improve application performance by running two copies of the computation, one on the mobile computer and another on a surrogate. Both copies start from the same initial state and receive the same inputs. Deterministic replay guarantees that both executions, if correct, produce identical output. Thus, as soon as either execution produces an external output (e.g., network data, screen message, etc.), that output is used immediately. However, as with Slingshot, a subsequent comparison may determine that the surrogate output differed from the local output; this indicates a faulty or malicious surrogate.

Compared to running only a single replica of an operation, deterministic replay does not require a prediction as to whether remote or local execution will be faster. Replay also guarantees that the state needed for each computation will be reproduced before the application begins that computation. Thus, there is no need to ship program state or a delta revision to the surrogate prior to each computation. Instead, the two computers need only exchange a log of non-deterministic events and their results. Further, this log can be sent asynchronously as events are generated, as opposed to a state delta which typically is generated and sent synchronously. This allows the surrogate to execute ahead during compute-intensive portions of an application, while the mobile computer can execute ahead during I/O-intensive portions. The user sees the lowest latency of the two executions.

It is useful to consider what types of events would be in such a log. One source of non-determinism is user input. This typically arrives at a low rate, and so requires little log space. Another source of non-determinism is network input. If the surrogate is along the network path that connects the mobile computer to the Internet [Balan et al., 2002], the surrogate can act as a Internet proxy for the mobile computer's network communication with minimal overhead. Since network communication is seen by both parties in this scenario, the log of non-deterministic events can omit the actual data and just include a reference to the observed communication segment. For mobile computers that use more than one network simultaneously [Higgins et al., 2010], it may be best to run the surrogate in the cloud. Non-determinism from file system inputs can be eliminated by replicating file system data on the surrogate. Some sources of non-deterministic input are challenging to handle; for instance, a video camera may produce data at a high rate. For applications that use such data, this form of redundant execution may not make sense.

Other sources of non-determinism are scheduling decisions and signal/interrupt delivery. When the application uses a single processor or can be assumed to not contain data races, these events occur at low frequency, and thus consume little log space. Use of a deterministic scheduling algorithm [Devietti et al., 2009] in which both parties independently make identical scheduling decisions can avoid communicating these events entirely.

If the surrogate executes an application faster than the mobile computer, it must send output to be displayed on the mobile computer over the network. For most applications, output events will be relatively small. For instance, it has been shown that it is often faster and more energy-efficient to run a Web browser on a surrogate, ship the graphical result of the rendering of Web pages to the mobile computer, and display it there than it is to run the Web browser on the mobile computer [Kim et al., 2006].

Replay starts by shipping the current state of the application from the mobile computer to the surrogate. For a running application, this cost is roughly proportional to the size of its address space. However, the cost can be mitigated, especially for freshly started applications, by storing executables and dynamic libraries in a distributed storage system and passing these objects to the surrogate by reference rather than by value.

Both the mobile computer and the surrogate execute the application. When the surrogate reaches a point in the execution that requires a non-deterministic input, it obtains the result locally if possible. This is the case for network input (which the server sees before the mobile computer because it acts as a proxy) and possibly for file system input, if a replicated storage solution is used. For other sources of non-determinism such as user input, the surrogate sends a query to the mobile computer asking it to perform the operation immediately. In this case, the mobile computer does not re-execute the operation when it reaches the identical point in its own computation. Instead, it uses the result produced earlier for the surrogate. Using this method, the server can execute multiple operations ahead of the mobile execution, achieving interactive performance similar to that of pure single-copy offload.

A similar approach to enforcing determinism is used by the Paranoid Android system [Portokalidis et al., 2010]. Paranoid Android uses deterministic record and replay to re-execute an application running on the phone on a *security server* located in the cloud. In the trust model targeted by this system, the applications running on a mobile phone are not considered to be trustworthy, but the security of such applications can be verified by running additional analysis code. Like the home server in Slingshot, the cloud-based security server is trusted. The security server is assumed to have more resources than a mobile phone, and it is therefore able to execute security checks that would be too expensive to run on the phone.

The security server uses deterministic replay to execute the exact sequence of application instructions executed on the phone. However, the security server also adds additional checks in the form of analysis code that verifies security properties while the application executes. In order to guarantee deterministic replay, such checks must be read-only; they cannot perturb the state of the running application. The Paranoid Android authors implemented a virus scanner and a taint analysis framework as proofs of concept for this model. If the analysis code detects a security violation, the fault is reported asynchronously to the mobile phone. The user may then take corrective action.

4.5   DISCUSSION

This chapter began by enumerating the security and privacy properties desired by cyber foraging participants. Surrogate providers wish to prevent operations hosted on surrogates from performing malicious or harmful actions, and they wish to provide performance isolation to guard against denial of service attacks on their customers. Since these concerns are similar to those of the operators of multi-tenant cloud data centers, the techniques used in those environments, namely virtual machine isolation and resource quotas, should work equally well for surrogate owners.

However, the needs of mobile computer owners are more challenging to address. The user of a surrogate wishes to ensure that the surrogate provides the correct results for a remote operation and that the surrogate does not divulge confidential information. Secure boot and verification via redundant execution providing promising but incomplete solutions to these problems. Secure boot does not guard against hardware attacks, physical tampering, or compromise of trusted software after it has been loaded on the surrogate. Verification does not provide confidentiality, and it has proven challenging to provide timely notification of incorrect results and to verify complex non-deterministic applications.

As this discussion indicates, the most pressing need for security and privacy research in cyber foraging systems is to develop better mechanisms for addressing the needs of mobile computer owners. For instance, the cloudlets vision posits that reputation-based trust mechanisms, in which the user verifies the identity of a surrogate provider and then relies on legal, business or other external mechanisms to establish trust, might be a minimally intrusive mechanism to enhance cyber foraging security. As observed in the Paranoid Android project, surrogates often have greater resources than mobile computers, so offloading expensive security mechanisms to a surrogate could potentially even enhance mobile security by enabling more powerful checks.

CHAPTER 5

Data Staging

The previous chapters in this lecture have focused on the use of surrogates for offloading computation from mobile computers. However, the original visions for cyber foraging [Balan et al., 2002; Satyanarayanan, 2001] posited a broader use of remote infrastructure in which surrogates also improve data transfers between mobile computers and the cloud by temporarily staging data in transit. This chapter explores this second use case.

Section 5.1 begins by outlining the motivation for employing data staging. By temporarily storing data sent between the cloud and mobile computers, a surrogate can enable the mobile computer to transfer data over a high bandwidth, one-hop wireless link rather than use a wide-area connection. This, in turn, can reduce the energy usage of the mobile computer as well as cellular bandwidth consumption. Section 5.2 describes several systems that speculatively prefetch data items to a surrogate so that they may later be retrieved by a mobile computer when it passes nearby. To be effective, such systems much predict the future; for instance, the data that the mobile computer will require, the path of the mobile computer's user, and the bandwidth available along that path. Section 5.3 details systems that provide such predictions. Section 5.4 describes a final use case in which surrogates stage data being sent in the opposite direction: from the mobile computer to the cloud. Finally, Section 5.5 summarizes and concludes the chapter.

5.1 MOTIVATION

In an age of near-ubiquitous connectivity, it may at first seem odd to employ surrogates to temporarily stage data sent between the cloud and mobile computers. However, if one carefully examines the capabilities of both wireless and wired networks located at the edge of the Internet, then it becomes apparent that such an architecture can offer substantial benefits.

Short range, one-hop networks such as those provided by WiFi access points typically offer both low latency and high bandwidth. The bandwidth shared between clients over the wireless network may often be an order of magnitude or more greater than the shared bandwidth provided by the next-hop, *backhaul* link that connects the wireless base station to the Internet [Dogar et al., 2010]. Since the vast majority of network traffic on the wireless network is between mobile computers and the cloud, the bottleneck for such traffic is the backhaul connection rather than the wireless link.

Now consider a mobile computer that passes near a WiFi access point for a short period of time. The applications on the computer may wish to prefetch data such as music, maps, news, and mail that the user is likely to access in the future. These applications initiate connections to data sources in the cloud and fetch as much data as possible while the computer is in range of the access point. As observed in the previous paragraph, the amount of data that is prefetched will often be limited by the backhaul connection rather than by the wireless network. In other words, the mobile computer cannot utilize the full bandwidth offered by the wireless network to prefetch data.

In contrast, the mobile computer can potentially prefetch significantly more data if a surrogate stages data on its behalf. Prior to the arrival of the mobile computer, the surrogate can prefetch data that the mobile computer might request from cloud sources over the limited backhaul network; the surrogate caches such data on its local storage. When the mobile computer passes nearby, it fetches the data from the surrogate using the full available bandwidth of the wireless, one-hop network. Given current disparities in network capacity between WiFi networks and backhaul connections, the mobile computer may potentially be able to fetch an order of magnitude more data when a surrogate provides data staging in this manner.

However, successful staging requires predicting the future: the surrogate must anticipate the arrival of the mobile computer and the set of data that it will request. Tools for making such predictions are described in Section 5.3.

While the discussion so far has focused on data fetched over WiFi connections, many mobile computers currently use cellular connections for data transfers. The ubiquity of cellular networks make them attractive for interactive data retrieval; however, such networks typically offer substantially lower bandwidth than even backhaul connections, making them a poorer option for prefetching.

In addition to offering greater throughput, staging data transfers from the cloud to the mobile computer can have substantial energy benefits as compared to unstaged transfers over WiFi and cellular networks. Characterization studies of wireless network power usage [Anand et al., 2003; Zhang et al., 2010] have shown that WiFi data transmission is typically much more energy-efficient than cellular data transmission for transmitting bulk data. Although WiFi interfaces often require more power to send and receive data, WiFi networks have substantially higher data transfer rates than cellular networks. Thus, when calculated as the cost per bit, the energy required for WiFi network transmission is lower than that of cellular network transmission in most cases.

Furthermore, even when considering only WiFi networks, research has shown that transmitting data as fast as possible is more energy efficient than sending data at slower speeds. Network interfaces, like many hardware devices, support power saving modes in which most energy-consuming components are disabled when the device is idle. However, the interface must be in a high-power mode to send and receive data. Additionally, transitioning into or out of the high-power mode takes time. Since the arrival of network packets tends to be correlated in time, a network packet often arrives soon after the processing of the previous packet has finished. A power management strategy that puts the device in a low-power mode as soon as it becomes idle can therefore introduce substantial delays; as soon as the device starts to enter a low-power mode, it must be woken up again to service a newly arrived packet.

For this reason, network power management strategies keep the device in a high-power mode for some time after transmitting a packet; transition to low-power mode only occurs after a timeout.

For most interfaces, the dominant energy cost is not the extra energy required to send and receive data, it is in fact the energy spent waiting in high-power mode to see if additional packets arrive. Therefore, the best data transmission strategy is simply to transmit data as fast as possible, so that the device can return to a low-power mode as soon as possible.

Many systems use the above observations to save energy. For instance, the Catnap system [Dogar et al., 2010] buffers data at a base station for a short period of time so that the entire buffer can be sent in a burst to a nearby mobile computer. Data staging on surrogates enjoys the same benefits, but the potential savings are larger because the storage capacity of a surrogate is much larger than that of a base station.

Data staging can also allow mobile computers to be more cost efficient. It is now common for cellular carriers to cap data transmission and assess fees or other penalty for excessive network usage. Mobile computers can reduce their cellular network usage by offloading transmissions to WiFi when possible [Balasubramanian et al., 2010]. By allowing a mobile computer to transmit more data when it is located near a surrogate with a WiFi access point, data staging can reduce the amount of data that the mobile computer needs to send over cellular networks. Even when the set of data that a mobile computer cannot be predicted exactly, data staging allows a mobile computer to prefetch more data that might potentially be used. By prefetching and caching a higher percentage of the data that a user might access, the mobile computer can service more requests from interactive applications from its local storage and require less on-demand usage of cellular networks [Higgins et al., 2012].

5.2   DATA STAGING FOR PREFETCHING DATA

Most of the work that has explored how fixed infrastructure can stage data for mobile computers has concentrated on prefetching data from the cloud to the mobile computer. This scenario is attractive for several reasons. First, mobile computers typically receive more data than they transmit. Second, data staging is not well-suited for time-sensitive data transmission because it buffers data on surrogates; in contrast, prefetching can tolerate delays because it is, by definition, non-interactive. Finally, since prefetching systems must already predict the set of data that may be needed by the user, the existing predictions can be used to decide which data to stage at surrogates.

5.2.1   INFOSTATIONS

The Infostations project [Goodman et al., 1997; Iacono and Rose, 2002] was the earliest project to investigate staging data for mobile computers on fixed infrastructure. The authors argued that intermittent, high-speed wireless networks are an attractive alternative to ubiquitous low-speed networks. This vision matches typical modern wireless network coverage in which WiFi access points provide intermittent geographical coverage and cellular networks provide lower-speed, but more ubiquitous, coverage. The Infostation authors observed that since the wireless throughput provided near Infostations was expected to be much greater than the wired network throughput, Infostations should prefetch and cache information. Thus, Infostations serve as the data staging repositories in their work.

Noting that mobility prediction is critical for data staging, the Infostation architecture includes a cluster controller that manages a network of geographically dispersed Infostations. If the path of a user is known, the cluster controller sends data to the next Infostation on the path while the user is downloading information from the previous Infostation. If the path is not known, the cluster controller sends data to the Infostations that the user is most likely to visit in the future.

5.2.2   DATA STAGING

The Data Staging project [Flinn et al., 2003] refined this vision and adapted it to cyber foraging by considering how untrusted and lightly managed surrogates could provide functionality equivalent to Infostations. Data Staging targeted prefetching of data in distributed file systems; the prototype supported prefetching in the context of Coda [Kistler and Satyanarayanan, 1992] and later work also supported the Network File System [Callaghan et al., 1995].

In a typical motivating scenario for Data Staging, an interactive application running on a mobile computer accesses files stored in a distributed file system. The file system attempts to reduce access time by caching files on the mobile computer, but limited space, continuous updates, and imperfect prediction prevent it from caching all but a portion of the files that the user might potentially read. Consequently, many files must be fetched on-demand from the distant file server. The user experiences many frustrating delays because the application reads multiple files sequentially and reading each file incurs multiple network round-trips.

Data Staging interposes a proxy on the mobile computer that intercepts file system traffic. When the proxy observes that remote file accesses are incurring high latency, it finds a surrogate in the nearby network environment that is willing to provide extra storage capacity. The proxy registers with the surrogate and stages the set of files that the user is most likely to access in the future.

Staging is expedited by a *data pump* that executes on an idle computer (this may be in the cloud located near the file server). When the client proxy wishes to stage a file, it sends a message to the pump through a secure channel. The pump authenticates the message, reads the file from the file system, encrypts the file, and generates a cryptographic hash of the data. The pump transmits the encrypted file to the surrogate and sends the file key and hash to the mobile computer through the secure channel. When a staged file is read by the application, the proxy fetches the file from the nearby surrogate, decrypts it, and uses the hash to verify that the file has not been modified. Prefetching files to the surrogate decreases the number of high-latency, blocking file accesses and dramatically reduces the number of long delays experienced by the user.

Reads of unstaged data are serviced using the base file system protocol. The prefetching of files to the surrogate proceeds concurrently with file system traffic. After each file is staged on the surrogate, it becomes immediately available for use. Thus, as the number of staged files grows, the percentage of cache misses that need to be serviced by the distant file server decreases, leading to significant improvement in interactive application performance.

Surrogates used for Data Staging need not be trusted because end-to-end encryption guarantees data privacy and cryptographic hashes provide end-to-end data integrity. Data Staging does not explicitly defend against denial-of-service attacks that render surrogates unavailable, so a malicious surrogate may periodically refuse to provide requested files to a mobile computer. However, if a malicious surrogate performs significantly worse than expected, the mobile computer may abandon use of the surrogate without further cost because the surrogate contains only soft state. The mobile computer can always fetch any needed data directly from file servers in the cloud over the wide-area network.

Data Staging surrogates require little management. Most of the staging mechanism is provided via commodity software. For instance, the Apache Web server is used as the base system and all modifications required to support a file system protocol are implemented as CGI scripts. Since surrogates host only soft state, no critical information is lost if the surrogate is disrupted by power failure or a system crash. For example, modifications to file data made by mobile computers are not buffered on surrogates.

While Data Staging does not explicitly predict user mobility, it does introduce the concept of *roles* to predict future data accesses. A role is an explicit grouping of files that is associated with a high-level task commonly performed by a user. For example, a graduate student may create roles such as `thesis`, `coursework`, and `personal`. Files are mapped to one or more roles; the existence of a mapping denotes the likelihood that a file will be accessed when the user is performing a specific role. A user may manually create such mappings, or the Data Staging system can automatically learn the mappings by observing the set of files accessed. Data Staging users specify the set of roles they are currently performing (these are referred to as *active* roles). When the mobile computer discovers a surrogate, it creates a prefetch list that is the union of the files specified for all active roles. It stages files in order of priority until its storage quota on the surrogate is exceeded.

5.2.3   SULULA

Developing countries often have poor Internet connectivity, making data staging solutions especially attractive. Sulula [Reda et al., 2010] is a kiosk-based solution that uses data staging to improve interactive access to personal data.

A Sulula prototype has been deployed in Ethiopia, where typical shared dialup connections have effective bandwidth delivered to a user of around 10–15 Kb/s. Users who visit public kiosks to read e-mail, Web pages from RSS feeds, and other information that is not readily cached because it is unique to each user suffer substantial interactive delays while their data is fetched over a slow link. Sulula aims to reduce such delay by prefetching personal data before a user arrives.

Sulula users send SMS messages to kiosk systems with an approximate arrival time and a set of data to be prefetched on their behalf. The kiosk system performs admission control: if a prefetch request cannot be satisfied given resource constraints, it suggests alternate times at which the data could be available. Later, when users arrive at kiosks at their previously scheduled time, their prefetched data is waiting for them. Thus, Sulula eliminates much of the wait time for its users when they access their personal data.

Sulula introduces the notion of *channels*, which are service-specific adapters that follow the protocol necessary to prefetch data from different cloud providers. For instance, Sulula provides channels for email, news feeds, and simple HTML data. Channels are akin to device drivers in that they can translate generic prefetch requests into messages that obey the service-specific protocol required to retrieve data.

5.3   PREDICTING THE FUTURE

A data staging system must predict the future in order to be effective. Potentially, three types of predictions are required: mobility patterns, data access patterns, and network connectivity.

5.3.1   MOBILITY PREDICTION

Since data is prefetched by a surrogate prior to the arrival of the mobile computer, the data staging system must be able to predict the mobility pattern of the computer's user well enough to achieve sufficiently high confidence that the mobile computer will in fact pass nearby. If the confidence that the mobile computer will come near the surrogate is too low, then it does not make sense to spend limited backhaul bandwidth and surrogate storage resources prefetching data for that computer.

Infostations rely on the cluster controller to predict mobility. In certain scenarios, such as when the mobile computer is traveling along a highway, such predictions may be highly accurate. In urban scenarios, predictions may be less accurate. However, there is an extensive body of existing work on predicting the movements of cellular network users [Akyildiz and Wang, 2004; Bhattacharya and Das, 1999; Liang and Haas, 2003] that can potentially be applied in the data staging domain.

As an alternative to predicting mobility patterns in the computing infrastructure, each mobile computer can supply predictions of its own mobility. A user may supply this information explicitly, as is done in Sulula when the user sends an SMS to a specific kiosk provider. Alternatively, the mobile computer may have access to other information, such as a GPS route plan, from which accurate mobility predictions can be derived.

If mobility predictions are not provided by the user and cannot be derived from external context, the mobile computer may use a forecasting system such as BreadCrumbs [Nicholson and Noble, 2008] to predict its future locations. BreadCrumbs is based on the observation that people are creatures of habit: they tend to take the same paths every day. A mobile computer running BreadCrumbs tracks its own movements and creates a predictive model based on the observed data. This model is in effect a personal device-specific mobility prediction. As long as the user continues to follow similar patterns, the future movements of the mobile computer can be determined with reasonable accuracy.

BreadCrumbs tracks location as latitude and longitude coordinates, rounding all values to three decimal places. At the authors' location (Ann Arbor, Michigan), this resolves to an area 110x80 m in size, which matches reasonably well with the coverage of many WiFi access points.

Location data can be observed using GPS or derived from observation of environmental conditions such as nearby network infrastructure and signal strength [Schilit et al., 2003]. Based upon conclusions drawn from earlier work [Song et al., 2004], BreadCrumbs models mobility using a second-order Markov model, with fallback to a first-order model when the second-order model has no prediction. Each state in the second-order model consists of the mobile computer's current location and its previous location. Thus, the second-order model inherently distinguishes direction of travel: two states at the same current location would have different previous states when the user is traveling in different directions.

The mobile computer continually builds the personal mobility model from periodic location observations. When a new location becomes available, the transition probability from the previous state is updated. If the mobile computer remains stationary, the model reflects this as a self-loop back to the same state.

When a location prediction is needed, BreadCrumbs supplies possible future locations from the derived mobility model. The authors found that BreadCrumbs was over 60% accurate in predicting the next state but only slightly over 20% accurate in predicting the state six transitions in the future. This may not be sufficiently accurate to support data staging. However, the authors note that the trial period was only one week and that predictions should improve with more data collection. For the purpose of data staging, it may not be as important to predict the exact time at which the mobile computer will pass a surrogate location as it is to predict whether the mobile computer will pass by at all within a relatively long time window. Unfortunately, the BreadCrumbs system was not evaluated for such queries (although the accuracy would likely be higher than the numbers for an exact state match cited above). Given the current importance of location in mobile computing, it is likely that location measurement and prediction system will both continue to improve.

5.3.2   DATA ACCESS PREDICTION

The data staging system must also predict the set of data that the mobile computer will request when it is co-located with the surrogate. The mobile computer's requirements may change over time as it fetches some data on demand using wide-area networks and as the activity of the computer's user changes. Additionally, the useful lifetime of any prefetched data must be long enough so that the data does not become stale during the interval between initiating the prefetch of the data to the surrogate and the arrival of the mobile computer.

Systems such as Sulula ask the user to explicitly specify the set of data that should be prefetched, while systems such as Data Staging make such predictions automatically by observing the access patterns of the user and learning which files tend to be accessed together in particular contexts.

Data Staging attempts to predict accesses for only a single class of data, namely files stored in a distributed file system. While many algorithms have been developed to prefetch file accesses, most such algorithms are designed only to predict the next few accesses in order to hide disk latency [Amer et al., 2002; Kroeger and Long, 2001].

The prefetching algorithms that are most applicable to data staging are those that predict large sets of files to cache on a mobile computer for access when the computer is disconnected from the Internet. One such algorithm is SEER [Kuenning and Popek, 1997]. SEER observes file accesses to calculate the lifetime semantic distance between pairs of files. This distance between accesses to file A and B is defined to be 0 if file A has not been closed before file B is opened, and the number of intervening file opens (including the open of file B) otherwise. SEER calculates the geometric mean of the lifetime semantic distance for pairs of files, tweaks the distances with a number of heuristics, and then applies a clustering algorithm to partition the set of files such that the distance between files in the same partition is small. If one file in a cluster is accessed, then SEER prefetches the rest of the cluster; this is based on the assumption that files that files that have been accessed together in the past will continue to be accessed together in the future. Compared to the roles used in Data Staging, the SEER algorithm may produce clusters without semantic meaning to the user. On the other hand, SEER may detect patterns in data access of which the user is unaware.

Modern mobile computers execute many applications that access different types of data. Newsreaders, Web browsers, map applications, e-mail readers, and many other applications all prefetch data to the mobile computer to improve interactive performance. Unfortunately, the data access patterns for these applications may differ. This may make it difficult to apply a blanket prediction strategy to all data required by the mobile computer. In order to determine what data to prefetch for data staging, it may prove necessary to provide an application-specific interface [Higgins et al., 2012] so that each application can make separate predictions for its own type of data.

5.3.3   BANDWIDTH PREDICTION

Bandwidth predictions are helpful in determining the amount of data that should be prefetched. Roughly speaking, the amount of data that should be buffered on a surrogate is equal to the dwell time of the mobile computer within the wireless network co-located with the surrogate multiplied by the difference in available bandwidth between the wireless network and the backhaul connection. Prefetching less than this amount results in the mobile computer not being able to utilize the full bandwidth of the wireless network while it is co-located with the surrogate. Prefetching more than this amount results in the mobile computer being unable to retrieve all data buffered on a surrogate; this situation means that resources were wasted prefetching and caching the unconsumed data.

In addition to predicting mobility patterns, BreadCrumbs also predicts the wireless network bandwidth that will be available to a mobile computer in the future. When a computer discovers a new access point, BreadCrumbs connects to reference servers in the cloud via the access point and measures connection quality (downstream bandwidth, upstream bandwidth, and latency). BreadCrumbs associates test results with the geographic coordinates at which they were taken. On subsequent encounters, it retests access points with a small probability in order to adjust for changing conditions over time.

BreadCrumbs provides connectivity forecasts by first predicting the possible future locations of the mobile computer and then using the connectivity measurements to predict network conditions at each such locations. Other systems have built similar connectivity databases. For instance, WiFi-Reports [Pang et al., 2009] measures and predicts commercial access points and allows users to collaboratively share their measurements without compromising location privacy.

5.4 DATA STAGING FOR DATA SENT TO THE CLOUD

While the discussion so far has discussed staging data on surrogates as the data is transmitted from the cloud to mobile computers, it is also possible to stage data transmitted in the opposite direction. The mobile computer sends data over the high-bandwidth, local network to the surrogate when it passes nearby. The surrogate stores the data and later forwards it to cloud destinations over the slower backhaul connection. This process offers much the same benefits of full utilization of wireless network bandwidth, mobile computer energy savings, and reduction of cellular network usage that were discussed in Section 5.1 in the context of prefetching.

Fluid replication [Kim et al., 2002] is an example of a system that stages data sent from mobile computers to the cloud. Fluid replication supports updates to files in a distributed file system in which users should see a consistent view of data from any client of the file system. Normally, when file data in such a system is updated, the changes are sent to file servers, which then propagate the changes to other clients that read the data.

In Fluid Replication, however, clients send updates to nearby *WayStations* that store the changes, provide nearby clients with updated file versions, and periodically propagate the changes back to the file servers. It is natural to consider providing WayStation-like functionality via cyber foraging surrogates. Mobile clients can propagate updates to a WayStation on a nearby surrogate much faster and with less energy expended than they can propagate those updates to file servers. The WayStation provides safety through replication: even if the mobile computer is lost, stolen, or damaged, the WayStation can still persist updates at the file server on its behalf. In the common case, the mobile computer will not need to transmit data to servers over wide-area networks.

Although Fluid Replication assumes that WayStations are reliable, it is possible to adapt the system design to have mobile computers retain a copy of all updates until they receive notification of the receipt of the update from a file server. With this change, a mobile computer could potentially detect surrogate failure with a time-out or an end-to-end cryptographic check for data corruption. Since surrogates would only retain soft state, a surrogate failure would not compromise data safety.

The techniques of Fluid Replication could potentially be applied to any distributed storage system with cloud servers and poorly connected clients. For instance, Horatio [Smaldone et al., 2009] provides similar functionality to Way Stations, but targets transmission of virtual machine images that are suspended and resumed on different clients.

Staging data in transit to the cloud in this manner makes sense when the transmission of data is asynchronous and can tolerate some delay. The mobile computer may need to buffer data prior to arrival at the surrogate, and the limited capacity of the backhaul link from the surrogate to the cloud may introduce further delays. Nevertheless, e-mail, file updates, and other non-interactive applications could potentially benefit from this use of cyber foraging.

5.5   SUMMARY

This chapter discussed how surrogates can assist mobile computers by staging data sent between those computers and the cloud. We began by listing three advantages of staging. First, staging allows mobile computers to send and receive data using all available bandwidth of the one-hop, wireless network rather than be bottlenecked by the more limited backhaul connection. Second, exchanging data with a surrogate is more energy-efficient than exchanging data with a cloud server. Finally, staging reduces cellular network usage, which may result in cost savings for the mobile user.

We next described several instances of data staging. We described three prefetching systems that stage data in transit from the cloud to the mobile computer: Infostations, Data Staging, and Sulula. Successful prefetching requires predicting the future. For data staging, at least three such predictions are required: where the mobile computer will be, what data it will require, and the network connectivity that it will have available along its route. We also described a system, Fluid Replication, that stages data traveling in the opposite direction: from the mobile computer to the cloud. All applications of data staging are most suitable for data transmissions that can tolerate some delay. In contrast, the store-and-forward nature of data staging will often be a poor match for interactive traffic.

Compared to computational offloading, data staging introduces less security and privacy concerns for surrogates. End-to-end encryption can provide privacy and tamper-resistance for data. If malicious or faulty surrogates deny service by refusing to forward data, mobile computers can initiate direct connections to the cloud and exchange data directly. As the Data Staging project showed, the functionality required to stage data is simple enough that it can be provided by standard software such as a Web server with a minimum of customization through CGI scripts. This enables staging surrogates to be packaged as embedded systems with minimal configuration, in much the same way that WiFi routers are currently packaged.

CHAPTER 6

Challenges and Opportunities

Cloud and mobile computing are dominant trends in modern technology. It is therefore natural to ask how one should best build systems that combine the two concepts. Current solutions tend to gravitate to one extreme or the other; either they place most functionality in the cloud via thin-client architectures, or they place most functionality on mobile devices via thick-client architectures.

The few current mobile applications that attempt to divide functionality between both endpoints statically partition data and computation. As this lecture has argued, static partitioning between cloud and mobile computers is poorly suited for mobile environments, in which resource availability changes rapidly.

Cyber foraging introduces two important innovations. First, in a cyber foraging system, the partition of functionality is dynamic; the mobile application adapts to changing resource availability by relocating data and computation to improve performance, reduce energy usage, or satisfy other high-level user goals. Second, cyber foraging systems opportunistically discover and use fixed infrastructure in the form of surrogate computers located close to the mobile computer's current location. Surrogates represent an intermediate tier between the mobile and cloud endpoints. Surrogates offer greater resources than what can be supplied by a mobile device, and they are reachable via higher quality networks than the wide-area links than those must be traversed to connect mobile computers and the cloud.

This chapter concludes the lecture by assessing the remaining barriers to widespread deployment of cyber foraging systems and outlining a research agenda that can help overcome those barriers. This discussion begins in the next section with an examination of modern computing trends and their impact on the future cost and benefit of cyber foraging. Section 6.2 lists the most pressing challenges that must be addressed before cyber foraging can be widely adopted. Section 6.3 describes a potential research agenda that may help address those challenges. Section 6.4 concludes with some final thoughts.

6.1 TRENDS THAT IMPACT CYBER FORAGING

In the introduction to this lecture, we argued that cyber foraging currently provide three substantial benefits to mobile computer users: improved performance, extended battery lifetime, and improved application fidelity. It is reasonable to consider whether these benefits are sustainable given current computing trends — in other words, are these benefits likely to increase, decrease, or stay the same in the future?

One alternative to cyber foraging is to perform all or most functionality on mobile computers. The storage and computational capacity of mobile devices will likely continue to increase, and such hardware advances will increase the speed at which current applications execute. On the other hand, mobility requires portability. Size, weight, and power constraints have always severely constrained the resources that can be offered on mobile platforms. It is unlikely that such constraints will be eliminated in the future. As Figure 1.1 showed, these constraints have caused the processing power of mobile computers to consistently lag behind that of servers.

The reason for this lag is that the computational and storage capacity of computers in fixed infrastructure deployments such as data centers has also been increasing rapidly. Since such deployments do not share the constraints of mobile computers, they are more easily scaled out (for example, by building more or larger data centers that host more computers). Widespread deployment of surrogates is another form of scaling out infrastructure capacity. Thus, fixed infrastructure can increase capacity through *both* hardware advances and scale out, meaning that the gap between mobile and infrastructure resource capacity is more likely to increase than decrease in the future.

The previous discussion begs the opposite question: perhaps the need for cyber foraging could be eliminated by executing functionality entirely or mostly in cloud data centers? Such execution necessitates communication since the mobile user interacting with applications is located far from the data center. Although cellular carriers are currently straining to deal with increased mobile data traffic, both the coverage and bandwidth provided by wireless networks have improved substantially in recent years. However, communication latency has not improved. While it may be feasible to shave some time from round-trip latencies by eliminating middleboxes and making routing more efficient, speed-of-light delays comprise the majority of round-trip latency for most wide-are networks. Cloud service providers have addressed latency concerns by replicating functionality across geographically distributed data centers; in effect, bringing the data center closer to the mobile user. Further distribution of data center functionality could reduce round-trip latency. However, the logical extension of this trend is the deployment of surrogates (i.e., cyber foraging).

Energy constraints are a fundamental concern for any battery-powered device. Unfortunately, battery capacity has improved at a slower rate than the improvement rate of processor speeds, storage capacity, or network bandwidth. This disparity has led to a noticeable gap: application complexity, and hence application demand for resources, has grown faster than the ability of mobile computer batteries to power those resources. Mobile system developers have filled the gap by deploying more sophisticated power management strategies that reduce energy wasted by hardware components when they are not performing tasks on behalf of the application. Over time, however, the power management strategies have had to introduce greater and greater complexity to eliminate ever smaller inefficiencies. This means that battery lifetime is likely to take on even greater importance as a limiting factor in mobile computing. Since cyber foraging offers the ability to shift resource usage from the mobile computer to fixed infrastructure, it provides a method for reducing mobile computer energy usage that is orthogonal to hardware power management. Thus, energy management is likely to be an even more compelling motivation for cyber foraging in the future.

The technology trends that are rapidly increasing the computing and storage capacity of mobile computers will likely make it feasible in the future to execute many applications on mobile devices that are simply too resource-hungry for mobile computers today. It is therefore possible that the third motivation for cyber foraging, namely the inability to execute demanding applications at full fidelity, will decrease in importance in the future. However, the resources demanded by applications have tended to swell over time with computing capacity. It is more likely, therefore, that new, even more resource-hungry applications will emerge. Such high-end applications may run well on fixed infrastructure, but not on less capable mobile devices. Thus, the particular applications that can use cyber foraging to provide increased fidelity may change over time, but a set of such applications is likely to always exist.

6.2   CHALLENGES

Given that current technology and future trends favor the deployment of cyber foraging systems, what are the most substantial barriers that might prevent such deployments from being successful?

6.2.1   FINDING THE KILLER APP

It is hard to identify one important application for which the benefits of cyber foraging are currently so compelling as to drive the deployment of cyber foraging infrastructure. Applications that benefit most from cyber foraging are both interactive and resource-intensive. Non-interactive applications are most easily executed in cloud data centers since such applications can, almost by definition, tolerate the additional latency required for communication between the mobile device and the cloud. Applications that are not resource-intensive can be executed on the mobile computer.

The applications that are currently most popular with mobile users (simple Web browsing, e-mail, and social applications) often require few local resources from the mobile computer. Consequently, current cyber foraging systems provide some benefits but are not especially compelling for these applications. On the other hand, the list of applications supported by current cyber foraging systems in Chapter 2 includes more demanding applications such as speech recognition, face recognition, and music identification that are increasingly popular with mobile users. The challenge for such applications will be demonstrating that the reduced latency of surrogate execution can provide noticeable usability improvements over execution in the cloud (or on the mobile device).

6.2.2   DEFINING THE BUSINESS CASE

It is unclear at this time what business forces will lead to the widespread deployment of surrogates. Partially, this is an issue of critical mass. Without cyber foraging applications, there is little economic demand for surrogates. Yet, without surrogates, there is little incentive to commercially develop cyber foraging systems and applications.

For this reason, it seems most likely that the development of cyber foraging applications and the deployment of surrogates will have to happen simultaneously. The most likely motivation is the need to deliver better performance for existing applications that statically partition their functionality between mobile phones and the cloud.

One possible avenue for deployment is for entities that host servers in the cloud to deploy surrogates as an extension of the data center. Cloud providers are currently deploying data centers in different parts of the world to improve network connectivity (e.g., latency) between data centers and the clients that access them. Content distribution networks achieve even greater geographic dispersal by caching content in many different locations. The logical extension of this trend would be to push the data center functionality as close to clients as possible. In this model, surrogates may remain centrally managed over the network and within the administrative domain of a cloud data center provider. Yet, geographic co-location with network access points would provide high-quality connectivity between surrogates and mobile computers.

Another possible avenue for deployment is bottom-up organic growth. In this model, current providers of WiFi hotspots may choose to augment their services by providing surrogates that caches data and hosts computation. As the price of computer hardware drops, the cost of deploying a surrogate may eventually be comparable to the cost of deploying a hotspot (especially after charges for backhaul connectivity are considered). The challenge for such organic deployment is management: the surrogate should be no harder to operate than a wireless hotspot.

A final possible avenue for deployment may be to have surrogates originally perform simpler functionality such as data staging. After such surrogates are deployed, they can evolve to perform more complex operations such as hosting remote computation. Cellular network operators, who are currently straining to handle the increased load of data-intensive mobile applications, have the incentive to deploy data-staging surrogates. If such surrogates can reduce cellular network load and/or make such load more predictable, then surrogates may be less costly to deploy than additional cellular network infrastructure.

6.2.3   SECURITY AND PRIVACY

Mobile computer users should naturally be uncomfortable with the concept of storing personal data or performing sensitive computation on computer systems outside of their immediate control. As we discussed in Chapter 4, existing approaches for providing security and privacy guarantees in cyber foraging systems are incomplete. More substantial work in this area is clearly required before commercial cyber foraging infrastructure can be deployed.

Trust will depend on the manner in which cyber foraging infrastructure is deployed. If surrogates are managed by a party such as a cloud service provider or a cellular network operator with whom the mobile user already has a trust relationship, then that user may be more willing to trust surrogates with sensitive data and computation. However, since poor surrogate security would then reflect badly on operators, third parties may be unwilling to deploy surrogates without substantial security and privacy guarantees that address, for example, the possibility of remote tampering. On the other hand, without substantially improved security and privacy mechanisms, mobile users may simply refuse to use surrogates provided by entities with whom they do not have a pre-existing trust relationship.

One possible avenue for deploying secure and private cyber foraging infrastructure is to initially limit the functionality provided by surrogates. For example, Chapter 5 argued that it is much easier to securely and privately stage data than it is to provide equivalent guarantees for remotely hosted computation. Hence, staging surrogates might help develop initial trust in the surrogate model from both users and operators. Surrogate functionality could be expanded gradually, for example to handle only remote computation that is known to be of little sensitivity, as robust security and privacy mechanisms are developed.

6.3   OPPORTUNITIES FOR FURTHER RESEARCH

Cyber foraging currently presents a rich set of opportunities for further research. Some of these opportunities will come from addressing the challenges listed in the previous section, while still others are the logical next step in the cyber foraging research described by this lecture.

6.3.1   BRINGING BENEFIT TO MORE APPLICATIONS

As discussed in Chapter 2, it is currently difficult for applications that do not have large chunks of computation between user interactions to benefit from cyber foraging. If a block of uninterrupted computation is small, then the time and energy saved by performing the computation remotely is less than the time and energy used to synchronously communicate inputs to the surrogate and receive back the results of the computation. This, unfortunately, often precludes many popular interactive applications such as Web browsers from benefiting substantially from cyber foraging.

A promising opportunity for further research is to develop new models of executing computation on surrogates that remove the need for synchronous communication. One possibility is to take advantage of natural parallelism in the application to overlap communication of inputs and results for one thread with the computation performed for other threads. Another possibility may be to predict what inputs will be needed before execution of a computation and ship those inputs to the surrogate in advance. Static analysis or profiling of application code could potentially help make such predictions more accurate. Alternatively, deterministic redundant execution on surrogates of the computation that generates inputs may be used in lieu of actually shipping those inputs. If one or more of the above methods can be developed to substantially lower the amount of synchronous communication required to execute portions of interactive applications on surrogates, then many applications such as Web browsers may be able to benefit from cyber foraging to a much greater extent than they can benefit today.

6.3.2   HANDLING UNCERTAINTY AND FAILURE

Chapter 2 discussed how cyber foraging systems currently have only rudimentary support for handling failures and severe performance degradations. For instance, several systems rely on timeouts to detect that a network or surrogate has failed.

One promising research direction that can address both failures and unpredictable performance is replication. A cyber foraging system can improve both average and worst-case performance by executing the same computation or storing data at multiple locations. For instance, it could run the computation on two surrogates, or it could execute the computation both on a surrogate and on the mobile computer. The application can proceed with the first result generated. Remaining executions can be aborted and their results discarded.

Executing multiple copies of a computation consumes extra resources (CPU cycles, network bandwidth, and energy), so a cyber foraging system may need to be judicious in deciding when to employ redundancy. One possibility is to execute multiple computations only when there is considerable uncertainty as to which location would yield the fastest response. Another possibility is to start a redundant computation only when an existing computation is taking longer than expected to return a result.

6.3.3   IMPROVING EASE-OF-MANAGEMENT

The potential business cases for deploying cyber foraging outlined in Section 6.2.2 require that surrogates be very easy to manage. One promising research direction is to develop software architectures that enable most problems to be solved by restarting software components on the surrogate [Candea et al., 2004]. Since many cyber foraging systems already employ strong isolation on surrogates in the form of virtual machines and eschew storing hard state on surrogates, such microreboot techniques may be more applicable on surrogates than on traditional computer systems. Additionally, research is needed to develop better tools that can remotely administer surrogates and diagnose software failures.

Surrogate management can also be simplified with research that improves security and privacy. Although trusted boot and verification are promising techniques, both leave surrogates vulnerable to physical attacks in which a determined eavesdropper modifies the surrogate hardware to snoop on data and communication. Additional research could focus on attacks that are most dangerous for computers in public locations.

Finally, the most significant issues that will come in the area of ease-of-management may not even become clear until there exists a small-scale surrogate deployment. Thus, developing a prototype or testbed that has an active user base should be a prerequisite for further research in this area.

6.4   FINAL THOUGHTS

Cyber foraging is still a research concept. There has yet to be a commercial cyber foraging deployment. Perhaps this is due to the remaining challenges that have yet to be addressed by research, the lack of a killer application, or the need to formulate a compelling business case.

However, the need for cyber foraging appears to be fundamental. Cyber foraging stands at the confluence of two strong trends in modern community: every day more people access data and run computation from their mobile computers, and data and computation is increasingly distributed among those devices and cloud data centers. Cyber foraging is especially compelling in a post-PC world in which desktop and workstation computers are replaced by mobile and cloud devices. The large variation in mobile computing environments argues that functionality should by dynamically, rather than statically, partitioned among mobile and cloud computers. Further, the limitations of locating computation in both cloud data centers (e.g., communication latency) and on mobile devices (e.g., limited resources) create the need to insert a surrogate computational tier between these extremes that can mitigate both sets of limitations. The next few years should bring many interesting developments to cyber foraging, as well as to the general combination of mobile and cloud computing.

Bibliography

Ian F. Akyildiz and Wenye Wang. The predictive user mobility profile framework for wireless multimedia networks. *IEEE/ACM Transactions on Networking*, 12(6):1021–1035, 2004. ISSN 1063-6692. DOI: 10.1109/TNET.2004.838604 Cited on page(s) 70

Ahmed Amer, Darrell D.E. Long, and Randal Burns. Group-based management of distributed file caches. In *Proceedings of the 22nd International Conference on Distributed Computing Systems*, Vienna, Austria, July 2002. DOI: 10.1109/ICDCS.2002.1022302 Cited on page(s) 71

Manish Anand, Edmund B. Nightingale, and Jason Flinn. Self-tuning wireless network power management. In *Proceedings of the 9th Annual Conference on Mobile Computing and Networking*, pages 176–189, San Diego, CA, September 2003. DOI: 10.1145/938985.939004 Cited on page(s) 19, 20, 66

Rajesh Balan. *Simplifying Cyber Foraging*. PhD thesis, Department of Computer Science, Carnegie Mellon University, May 2006. Cited on page(s) 22

Rajesh Balan, Jason Flinn, M. Satyanarayanan, Shafeeq Sinnamohideen, and Hen-I Yang. The case for cyber foraging. In *the 10th ACM SIGOPS European Workshop*, Saint-Emilion, France, September 2002. DOI: 10.1145/1133373.1133390 Cited on page(s) 62, 65

Rajesh K. Balan, Mahadev Satyanarayanan, So Young Park, and Tadashi Okoshi. Tactics-based remote execution for mobile computing. In *Proceedings of the 1st International Conference on Mobile Systems, Applications and Services*, pages 273–286, San Francisco, CA, May 2003. DOI: 10.1145/1066116.1066125 Cited on page(s) 8, 9, 10, 24

Rajesh Krishna Balan, Darren Gergle, Mahadev Satyanarayanan, and James Herbsleb. Simplifying cyber foraging for mobile devices. In *Proceedings of the 5th International Conference on Mobile Systems, Applications and Services*, San Juan, Puerto Rico, June 2007. DOI: 10.1145/1247660.1247692 Cited on page(s) 10, 12

Aruna Balasubramanian, Ratul Mahajan, and Arun Venkataramani. Augmenting mobile 3G using WiFi. In *Proceedings of the 8th International Conference on Mobile Systems, Applications and Services*, pages 123–136, San Francisco, CA, June 2010. DOI: 10.1145/1814433.1814456 Cited on page(s) 20, 67

Paul Barham, Boris Dragovic, Keir Fraser, Steven Hand, and Tim Harris. Xen and the art of virtualization. In *Proceedings of the 19th ACM Symposium on Operating Systems Principles*, pages

164–177, Bolton Landing, NY, October 2003. DOI: 10.1145/945445.945462 Cited on page(s) 37

Fabrice Bellard. QEMU: A fast and portable dynamic translator. In *FREENIX Track: Proceedings of the USENIX Annual Technical Conference*, pages 41–46, April 2005. Cited on page(s) 36

Amiya Bhattacharya and Sajal K. Das. Lezi-update: An information-theoretic approach to track mobile users in PCS networks. In *Proceedings of the 5th International Conference on Mobile Computing and Networking*, pages 1–12, Seattle, WA, August 1999. DOI: 10.1145/313451.313457 Cited on page(s) 70

Aaron B. Brown and David A. Patterson. Rewind, repair, replay: Three R's to dependability. In *the 10th ACM SIGOPS European Workshop*, St. Emilion, France, September 2002. DOI: 10.1145/1133373.1133387 Cited on page(s) 60

Brad Calder, Ju Wang, Aaron Ogus, Niranjan Nilakantan, Arild Skolsvold, Sam McKelvie, Yikang Xu, Shashwat Sriastav, Jiesheng Wu, Huseyin Simitci, Jaidev Haridas, Chakravarthy Uddaraju, Hemal Khatri, Andrew Edwards, Vaman Bedekar, Shane Mainali, Rafay Abbasi, Arpit Agarwal, Mian Fahim ul Haq, Muhammed Ikran ul Haq, Deepali Bhardwaj, Sowmuya Dayanard, Anitha Adusumilli, Marvin McNett, Sriram Sankaran, Kavitha Manivannan, and Leonidas Rigas. Windows Azure Storage: A highly available cloud storage with strong consistency. In *Proceedings of the 23rd ACM Symposium on Operating Systems Principles*, Cascais, Portugal, October 2011. DOI: 10.1145/2043556.2043571 Cited on page(s) 55

B. Callaghan, B. Pawlowski, and P. Staubach. NFS Version 3 Protocol Specification. Technical Report RFC 1813, IETF, June 1995. Cited on page(s) 68

George Candea, Shinichi Kawamoto, Yuichi Fujiki, Greg Friedman, and Armando Fox. Microreboot – A technique for cheap recovery. In *Proceedings of the 6th Symposium on Operating Systems Design and Implementation*, pages 31–44, San Francisco, CA, December 2004. Cited on page(s) 80

Byung-Gon Chun, Sunghwan Ihm, Petros Maniatis, Mayur Naik, and Ashwin Patti. CloneCloud: Elastic execution between mobile device and cloud. In *Proceedings of the European Conference on Computer Systems*, Salzburg, Austria, April 2011. DOI: 10.1145/1966445.1966473 Cited on page(s) 7, 8, 10, 32

Asaf Cidon, Tomer M. London, Sachin Katti, Christos Kozyrakis, and Mendel Rosenblum. MARS: Adaptive remote execution for multi-threaded mobile devices. In *Proceedings of the 3rd ACM SOSP Workshop on Networking, Systems, and Applications on Mobile Handhelds (MobiHeld)*, Cascais, Portugal, October 2011. DOI: 10.1145/2043106.2043107 Cited on page(s) 15, 32

Landon P. Cox, Christopher D. Murray, and Brian D. Noble. Pastiche: Making backup cheap and easy. In *Proceedings of the 5th Symposium on Operating Systems Design and Implementation*, pages 285–298, Boston, MA, December 2002. DOI: 10.1145/1060289.1060316 Cited on page(s) 45

Eduardo Cuervo, Aruna Balasubramanian, Dae ki Cho, Alec Wolman, Stefan Saroiu, Ranveer Chandra, and Paramvir Bahl. MAUI: Making smartphones last longer with code offload. In *Proceedings of the 8th International Conference on Mobile Systems, Applications and Services*, pages 49–62, San Francisco, CA, June 2010. DOI: 10.1145/1814433.1814441 Cited on page(s) 8, 30

Dallas Semiconductor Corporation. *DS2437 Smart Battery Monitor*. 4401 South Beltwood Parkway, Dallas, TX, 1999. Cited on page(s) 21

Giuseppe DeCandia, Denix Hastorun, Madan Jampani, Gunavardhan Kakulapati, Avinash Lakshman, Alex Pilchin, Swaminathan Sivasubramanian, Peter Vosshall, and Werner Vogels. Dynamo: Amazon's highly available key-value store. In *Proceedings of the 21st ACM Symposium on Operating Systems Principles*, Stevenson, WA, October 2007. DOI: 10.1145/1323293.1294281 Cited on page(s) 30

Joseph Devietti, Brandon Lucia, Luis Ceze, and Mark Oskin. DMP: Deterministic shared memory multiprocessing. In *Proceedings of the 2009 International Conference on Architectural Support for Programming Languages and Operating Systems (ASPLOS)*, pages 85–96, March 2009. DOI: 10.1145/1508284.1508255 Cited on page(s) 63

Fahad R. Dogar, Peter Steenkiste, and Konstantina Papagiannaki. Catnap: Exploiting high bandwidth wireless interfaces to save energy for mobile devices. In *Proceedings of the 8th International Conference on Mobile Systems, Applications and Services*, San Francisco, CA, June 2010. DOI: 10.1145/1814433.1814446 Cited on page(s) 65, 67

E. N. Elnozahy, Lorenzo Alvisi, Yi-Min Wang, and David B. Johnson. A survey of rollback-recovery protocols in message-passing systems. *ACM Computing Surveys*, 34(3):375–408, September 2002. DOI: 10.1145/568522.568525 Cited on page(s) 23

William Enck, Peter Gilbert, Byung gon Chun, Landon P. Cox, Jaeyeon Jung, Patrick McDaniel, and Anmol N. Sheth. TaintDroid: An information-flow tracking system for realtime privacy monitoring on smartphones. In *Proceedings of the 9th Symposium on Operating Systems Design and Implementation*, Vancouver, BC, October 2010. Cited on page(s) 5

Jason Flinn and Z. Morley Mao. Can deterministic replay be an enabling tool for mobile computing? In *Proceedings of the 12th Workshop on Mobile Computing Systems and Applications (HotMobile)*, Phoenix, AZ, March 2011. DOI: 10.1145/2184489.2184507 Cited on page(s) 62

Jason Flinn and Mahadev Satyanarayanan. Managing battery lifetime with energy-aware adaptation. *ACM Transactions on Computer Systems (TOCS)*, 22(2):137–179, May 2004. DOI: 10.1145/986533.986534 Cited on page(s) 9

Jason Flinn, Keith I. Farkas, and Jennifer Anderson. *Power and Energy Characterization of the Itsy Pocket Computer (Version 1.5)*. Compaq Western Research Laboratory, February 2000. Technical Note TN-56. Cited on page(s) 21

Jason Flinn, SoYoung Park, and Mahadev Satyanarayanan. Balancing Performance, Energy, and Quality in Pervasive Computing. In *Proceedings of the 22nd International Conference on. Distributed Computing Systems*, Vienna, Austria, July 2002. DOI: 10.1109/ICDCS.2002.1022259 Cited on page(s) 8

Jason Flinn, Shafeeq Sinnamohideen, Niraj Tolia, and M. Satyanarayanan. Data staging for untrusted surrogates. In *Proceedings of the 2nd USENIX Conference on File and Storage Technologies*, pages 15–28, San Francisco, CA, March/April 2003. Cited on page(s) 68

Robert Frederking and Ralf D. Brown. The Pangloss-Lite machine translation system. In *Expanding MT Horizons: Proceedings of the Second Conference of the Association for Machine Translation in the Americas*, pages 268–272, Montreal, Canada, 1996. Cited on page(s) 10

Scott Garriss, Ramón Cáceres, Stefan Berger, Reiner Sailer, Leendert van Doorn, and Xiaolan Zhang. Trustworthy and personalized computing on public kiosks. In *Proceedings of the 6th International Conference on Mobile Systems, Applications and Services*, Breckenridge, CO, June 2008. DOI: 10.1145/1378600.1378623 Cited on page(s) 56, 58

Gartner. Gartner says consumers will spend 6.2 billion in mobile application store in 2010, January 2010. http://www.gartner.com/it/page.jsp?id=1282413. Cited on page(s) 1

David J. Goodman, Joan Borras, Narayan B. Mandayam, and Roy D. Yates. Infostations: A new system model for data and messaging services. In *Vehicular Technology Conference*, pages 969–973, 1997. DOI: 10.1109/VETEC.1997.600473 Cited on page(s) 67

Sachin Goyal and John Carter. A lightweight secure cyber foraging infrastructure for resource-constrained devices. In *Proceedings of the 6th IEEE Workshop on Mobile Computing Systems and Applications*, Lake Windermere, England, December 2004. DOI: 10.1109/MCSA.2004.2 Cited on page(s) 37, 55

Brett D. Higgins, Azarias Reda, Timur Alperovich, Jason Flinn, Thomas J. Giuli, Brian Noble, and David Watson. Intentional networking: Opportunistic exploitation of mobile network diversity. In *Proceedings of the 16th International Conference on Mobile Computing and Networking*, Chicago, IL, September 2010. DOI: 10.1145/1859995.1860005 Cited on page(s) 20, 63

Brett D. Higgins, Jason Flinn, Thomas J. Giuli, Brian Noble, Christopher Peplin, and David Watson. Informed mobile prefetching. In *Proceedings of the 10th International Conference on Mobile Systems, Applications and Services*, Low Wood Bay, England, June 2012. DOI: 10.1145/2307636.2307651 Cited on page(s) 9, 67, 72

Galen C. Hunt and Michael L. Scott. The Coign automatic distributed partitioning system. In *Proceedings of the 3rd Symposium on Operating Systems Design and Implementation*, pages 187–200, New Orleans, LA, February 1999. Cited on page(s) 11

Ana Iacono and Chrisopher Rose. Infostations: New perspectives on wireless data. *The Kluwer International Series in Engineering and Computer Science*, 598(3), 2002. DOI: 10.1007/0-306-47310-0_2 Cited on page(s) 67

IEEE Local and Metropolitan Area Network Standards Committee. *Wireless LAN medium access control (MAC) and physical layer (PHY) specifications.* New York, New York, 1997. IEEE Std 802.11-1997. Cited on page(s) 19

Intel and Microsoft. *Advanced Power Management (APM) BIOS Interface Specification*, February 1996. Cited on page(s) 20

Intel, Microsoft, and Toshiba. *Advanced Configuration and Power Interface Specification*, February 1998. http://www.acpi.info. Cited on page(s) 13, 20

Hyojun Kim, Nitin Agrawal, and Cristian Ungureanu. Examining storage performance on mobile devices. In *Proceedings of the 3rd ACM SOSP Workshop on Networking, Systems, and Applications on Mobile Handhelds (MobiHeld)*, Cascais, Portugal, October 2011. DOI: 10.1145/2043106.2043112 Cited on page(s) 22

Joeng Kim, Ricardo Baratto, and Jason Nieh. pTHINC: A thin-client architecture for mobile wireless web. In *Proceedings of the 15th International World Wide Web Conference (WWW 2006)*, 2006. DOI: 10.1145/1135777.1135803 Cited on page(s) 63

Minkyong Kim and Brian D. Noble. Mobile network estimation. In *Proceedings of the 7th International Conference on Mobile Computing and Networking*, pages 298–309, July 2001. DOI: 10.1145/381677.381705 Cited on page(s) 19

Minkyong Kim, Landon P. Cox, and Brian D. Noble. Safety, visibility, and performance in a wide-area file system. In *Proceedings of the 1st USENIX Conference on File and Storage Technologies*, Monterey, CA, January 2002. Cited on page(s) 73

Samuel T. King, Joseph Tucek, Anthony Cozzie, Chris Grier, Weihang Jiang, and Yuanyuan Zhou. Designing and implementing malicious hardware. In *Proceedings of the First USENIX Workshop on Large-Scale Exploits and Emergent Threats*, April 2008. Cited on page(s) 58

J. J. Kistler and M. Satyanarayanan. Disconnected operation in the Coda file system. *ACM Transactions on Computer Systems*, 10(1), February 1992. DOI: 10.1145/146941.146942 Cited on page(s) 22, 47, 68

M. Kozuch and M. Satyarayanan. Internet Suspend/Resume. In *Proceedings of the 4th IEEE Workshop on Mobile Computing Systems and Applications*, Callicoon, NY, June 2002. DOI: 10.1109/MCSA.2002.1017484 Cited on page(s) 37

Mads Daro Kristensen. Scavenger: Transparent deployment of efficient cyber foraging applications. In *IEEE International Conference on Pervasive Computing and Communications (PerCom)*, pages 217–226, San Diego, CA, March 2010. DOI: 10.1109/PERCOM.2010.5466972 Cited on page(s) 39

Thomas M. Kroeger and Darrell D.E. Long. Design and implementation of a predictive file prefetching algorithm. In *Proceedings of the 2001 USENIX Annual Technical Conference*, Boston, MA, June 2001. Cited on page(s) 71

Geoffrey H. Kuenning and Gerald J. Popek. Automated hoarding for mobile computers. In *Proceedings of the 21st ACM Symposium on Operating Systems Principles*, Saint-Malo, France, October 1997. DOI: 10.1145/268998.266706 Cited on page(s) 72

Jinyuan Li, Maxwell Krohn, David Mazières, and Dennis Shasha. Secure untrusted data repository (SUNDR). In *Proceedings of the 6th Symposium on Operating Systems Design and Implementation*, pages 121–136, San Francisco, CA, December 2004. Cited on page(s) 60

B. Liang and Z.J. Haas. Predictive distance-based mobility management for multidimensional PCS networks. *IEEE/ACM Transactions on Networking*, 11(5):718–732, October 2003. DOI: 10.1109/TNET.2003.815301 Cited on page(s) 70

Thomas L. Martin. *Balancing Batteries, Power, and Performance: System Issues in CPU Speed-Setting for Mobile Computing*. PhD thesis, Department of Electrical and Computer Engineering, Carnegie Mellon University, 1999. Cited on page(s) 21

Jonathan M. McCune, Adrian Perrig, and Michael K. Reiter. Bump in the ether: A framework for securing sensitive user input. In *Proceedings of the USENIX 2006 Annual Technical Conference*, Boston, MA, June 2006. Cited on page(s) 57

Dushyanth Narayanan. *Operating System Support for Mobile Interactive Applications*. PhD thesis, Department of Computer Science, Carnegie Mellon University, August 2002. Cited on page(s) 14

Dushyanth Narayanan, Jason Flinn, and Mahadev Satyanarayanan. Using history to improve mobile application adaptation. In *Proceedings of the 2nd IEEE Workshop on Mobile Computing Systems and Applications*, pages 30–41, Monterey, CA, August 2000. DOI: 10.1109/MCSA.2000.895379 Cited on page(s) 14, 15

David Nichols. Using idle workstations in a shared computing environment. In *Proceedings of the 11th ACM Symposium on Operating Systems Principles*, pages 5–12, Austin, TX, November 1987. DOI: 10.1145/41457.37502 Cited on page(s) 47

A. J. Nicholson and B. D. Noble. BreadCrumbs: Forecasting mobile connectivity. In *Proceedings of the 14th International Conference on Mobile Computing and Networking*, pages 46–57, San Francisco, CA, September 2008. DOI: 10.1145/1409944.1409952 Cited on page(s) 70

Brian D. Noble, M. Satyanarayanan, Dushyanth Narayanan, J. Eric Tilton, Jason Flinn, and Kevin R. Walker. Agile application-aware adaptation for mobility. In *Proceedings of the 16th ACM Symposium on Operating Systems Principles*, pages 276–287, Saint-Malo, France, October 1997. DOI: 10.1145/268998.266708 Cited on page(s) 18

Erik Nygren, Ramesh K. Sitaraman, and Jennifer Sun. The Akamai network: A platform for high-performance Internet applications. *ACM SIGOPS Operating Systems Review*, 44(3):2–19, July 2010. DOI: 10.1145/1842733.1842736 Cited on page(s) 30

Jefferey Pang, Ben Greenstein, Michael Kaminsky, Damon Mccoy, and Srinivasan Seshan. Wifi-reports: Improving wireless network selection with collaboration. In *Proceedings of the 7th International Conference on Mobile Systems, Applications and Services*, pages 123–136, Krakow, Poland, June 2009. DOI: 10.1145/1555816.1555830 Cited on page(s) 73

Padmanabhan S. Pillai, Lily B. Mummert, Steven W. Schlosser, Rahul Sukthankar, and Casey J. Helfrich. SLIPstream: Scalable low-latency interactive perception on streaming data. In *Proceedings of the ACM International Workshop on Network and Operating System Support for Digital Audio and Video*, Williamsburgy, VA, June 2009. DOI: 10.1145/1542245.1542256 Cited on page(s) 11

Georgios Portokalidis, Philip Homburg, Kostas Anagnostakis, and Herbert Bos. Paranoid android: Versatile protection for smartphones. In *Proceedings of the Annual Computer Security Applications Conference*, December 2010. DOI: 10.1145/1920261.1920313 Cited on page(s) 63

Moo-Ryong Ra, Anmol Sheth, Lily Mummert, Padmanabhan Pillai, David Wetherall, and Ramesh Govidan. Odessa: Enabling interactive perception applications on mobile devices. In *Proceedings of the 9th International Conference on Mobile Systems, Applications and Services*, Washington, DC, June 2011. DOI: 10.1145/1999995.2000000 Cited on page(s) 8, 18

Azarias Reda, Brian Noble, and Yidnekachew Haile. Distributing private data in challenged network environments. In *Proceedings of the Interntational World Wide Web Conference*, pages 801–810, Raleigh, NC, May 2010. DOI: 10.1145/1772690.1772772 Cited on page(s) 69

Rodrigo Rodrigues, Miguel Castro, and Barbara Liskov. BASE: Using abstraction to improve fault tolerance. In *Proceedings of the 18th ACM Symposium on Operating Systems Principles*, pages 15–28, Banff, Canada, October 2001. DOI: 10.1145/859716.859718 Cited on page(s) 61

Alexey Rudenko, Peter Reiher, Gerald J. Popek, and Geoffrey H. Kuenning. Saving portable computer battery power through remote process execution. *Mobile Computing and Communications Review*, 2(1):19–26, January 1998. DOI: 10.1145/584007.584008 Cited on page(s) 7, 8

Constantine P. Sapuntzakis, Ramesh Chandra, Ben Pfaff, Jim Chow, Monica S. Lam, and Mendel Rosenblum. Optimizing the migration of virtual computers. In *Proceedings of the 5th Symposium on Operating Systems Design and Implementation*, pages 377–390, Boston, MA, December 2002. DOI: 10.1145/1060289.1060324 Cited on page(s) 44

M. Satyanarayanan. Pervasive Computing: Vision and Challenges. *IEEE Personal Communications*, 8(4):10–17, August 2001. DOI: 10.1109/98.943998 Cited on page(s) 2, 29, 30, 32, 65

Mahadev Satyanarayanan, Paramvir Bahl, Ramón Cáceres, and Nigel Davies. The case for VM-based cloudlets in mobile comptuing. *IEEE Pervasive Computing*, 8(4):14–23, October–December 2009. DOI: 10.1109/MPRV.2009.82 Cited on page(s) 31, 37, 46, 58

Bill Schilit, Anthony LaMarca, Gaetano Borriello, William Griswold, David McDonald, Edward Lazowska, Anand Balachandran, and Vaughn Iverson. Challenge: Ubiquitous location-aware computing and the Place Lab initiative. In *Proceedings of the 1st ACM International Workshop on Wireless Mobile Applications and Services on WLAN Hotspots (WMASH)*, San Diego, CA, September 2003. DOI: 10.1145/941326.941331 Cited on page(s) 71

Fred B. Schneider. Implementing fault-tolerant services using the state machine approach: a tutorial. *ACM Computing Surveys*, 22(4):299–319, December 1990. DOI: 10.1145/98163.98167 Cited on page(s) 25

Aaron Schulman, Vishnu Ravda, Ramachandran Ramjee, Neil Spring, Pralhad Deshpande, Calvin Grunewald, Kamal Jain, and Venkata N. Padmanabhan. Bartendr: A practical approach to energy-aware cellular data scheduling. In *Proceedings of the 16th International Conference on Mobile Computing and Networking*, Chicago, IL, USA, September 2010. DOI: 10.1145/1859995.1860006 Cited on page(s) 20

Stephen Smaldone, Benjamin Gilbert, Nilton Bila, Liviu Iftode, Eyal de Lara, and Mahadev Satyanarayanan. Leveraging smart phones to reduce mobility footprints. In *Proceedings of the 7th International Conference on Mobile Systems, Applications and Services*, pages 109–122, Krakow, Poland, June 2009. DOI: 10.1145/1555816.1555828 Cited on page(s) 73

L. Song, D. Kotz, R. Jain, and X. He. Evaluating location predictors with extensive Wi-Fi mobility data. In *Proceedings of the 23rd Annual Joint Conference of the IEEE Computer and Communications Societies*, pages 1414–1424, March 2004. DOI: 10.1109/INFCOM.2004.1357026 Cited on page(s) 71

Joao Pedro Sousa, Rajesh Krishna Balan, Vahe Poladian, David Garlan, and Mahadev Satyanarayanan. User guidance of resource-adaptive systems. In *Proceedings of the 3rd International Conference on Software and Data Technologies*, July 2008. Cited on page(s) 9

Ya-Yunn Su and Jason Flinn. Slingshot: Deploying stateful services in wireless hotspots. In *Proceedings of the 3rd International Conference on Mobile Systems, Applications and Services*, pages 79–92, Seattle, WA, June 2005a. DOI: 10.1145/1067170.1067180 Cited on page(s) 24, 32, 36, 59

Ya-Yunn Su and Jason Flinn. Portable storage support for cyber-foraging. In *Proceedings of the 1st International Workshop on Software Support for Portable Storage*, pages 62–68, San Francisco, CA, March 2005b. Cited on page(s) 45

Niraj Tolia, Jan Harkes, Michael Kozuch, and M. Satyanarayanan. Integrating portable and distributed storage. In *Proceedings of the 3rd USENIX Conference on File and Storage Technologies*, San Francisco, CA, March/April 2004. Cited on page(s) 44

Andrew Whitaker, Marianne Shaw, and Steven D. Gribble. Scale and performance in the Denali isolation kernel. In *Proceedings of the 5th Symposium on Operating Systems Design and Implementation*, pages 195–209, Boston, MA, December 2002. DOI: 10.1145/1060289.1060308 Cited on page(s) 34

Peter Young. *Recrsive Estimation and Time-Series Analysis*. Springer-Verilag, Germany, 1984. DOI: 10.1007/978-3-642-82336-7 Cited on page(s) 15

Aydan R. Yumerefendi, Benjamin Mickle, and Landon P. Cox. TightLip: Keeping applications from spilling the beans. In *Proceedings of the 4th Symposium on Networked Systems Design and Implementation*, pages 159–172, Cambridge, MA, April 2007. Cited on page(s) 8

Heng Zeng, Carla S. Ellis, Alvin R. Lebeck, and Amin Vahdat. ECOSystem: Managing energy as a first class operating system resource. In *Proceedings of the 10th International Conference on Architectural Support for Programming Languages and Operating Systems (ASPLOS-X)*, San Jose, CA, October 2002. DOI: 10.1145/605397.605411 Cited on page(s) 21

Lide Zhang, Birjodh Tiawana, Zhiyun Qian, Zhaoguang Wang, Robert P. Dick, Zhuoqing Morley Mao, and Lei Yang. Accurate online power estimation and automatic battery behavior based power model generation for smartphones. In *Proceedings of the eighth IEEE/ACM/IFIP International Conference on Hardware/Software Codesign and System Synthesis*, pages 105–114, Scottsdale, AZ, USA, October 2010. DOI: 10.1145/1878961.1878982 Cited on page(s) 21, 66

Author's Biography

JASON FLINN

Jason Flinn is an Associate Professor of Computer Science and Engineering at the University of Michigan, Ann Arbor. He received his Ph.D. from Carnegie Mellon University in 2001. He has served as technical program committee co-chair of the International Conference on Mobile Systems, Applications, and Services (MobiSys) and the USENIX Conference on File and Storage Technologies (FAST). He received an NSF Career award and best paper awards from numerous conferences. His research interests include mobile computing, operating systems, storage, and distributed systems.

Printed in the United States
by Baker & Taylor Publisher Services